TVA
AND THE
GRASS ROOTS

A Study of Politics and Organization

BY
PHILIP SELZNICK

with 2011 Foreword by Jonathan Simon

Classics of the Social Sciences

Quid Pro Books
New Orleans, Louisiana

TVA AND THE GRASS ROOTS

Previously published in 1984 by the University of California Press (Berkeley and Los Angeles, California). Originally published as Volume 3, University of California Publications in Culture and Society, April 1, 1949, under the title *TVA and the Grass Roots: A Study in the Sociology of Formal Organization*.

Published in the 2011 *Classics of the Social Sciences* edition by Quid Pro Books.

QUID PRO, LLC
5860 Citrus Blvd., Suite D-101
New Orleans, Louisiana 70123
www.quidprobooks.com

ISBN: 161027055X (pbk, 2011)
ISBN-13: 9781610270557 (pbk, 2011)
ISBN: 1610270541 (Kindle)
ISBN: 1610270533 (ePub)

This new presentation of Philip Selznick's *TVA and the Grass Roots* is part of a series, *Classics of the Social Sciences*, which will include several of his early works. Proceeds from his books in the Series benefit the Philip Selznick Scholarship Fund in the Jurisprudence & Social Policy Program, University of California at Berkeley. This is an authorized edition.

The cover image is adapted from the original plans for the Norris Dam, built by the TVA on the Clinch River in Tennessee. The TVA adopted this design from a previous design that the U.S. Army Corps of Engineers had drafted for the "Cove Creek Project" in the late 1920s. These plans were created by a worker for the TVA under his employment by the U.S. government and are a public domain image.

Printed in the United States of America.

CONTENTS

[The original page numbers are embedded into text using {brackets}, for continuity of citation, syllabi, and referencing (across all prior authorized editions, 1949–1984). Cross-references refer to this original pagination. The footnote numbering is retained. The subject-matter Index is included, and likewise references the original pagination.

Page numbers in brackets below reference the original pagination. The unbracketed numbers to the right reference the new pagination.]

TABLES

CHARTS

[Page numbers above reference the actual location in this edition.]

2011 FOREWORD

TO STUDENTS OF formal organizations, whether in sociology, business, or politics, Philip Selznick's *TVA and the Grass Roots: A Study in the Sociology of Formal Organization*, first published in 1949 by the University of California, and then reissued as a paperback by Harper & Row in 1966, remains a classic example of studying a formal organization by investigating its informal structures. While other theoretical toolkits may have become more fashionable in the intervening decades (network analysis certainly, as well as public choice theory), *TVA* remains the template of using theory to guide in-depth empirical research on organizations, both to shape its research agenda and to unlock the insights in data (especially qualitative data). In Selznick's case the theory was coming from Robert Michels, Talcott Parsons, Robert Merton, Chester Bernard, and his own original efforts (it began as his doctoral dissertation). But compared to the abstractions of especially Parsons and Merton, *TVA and the Grass Roots* shows theory at its very best use, opening up fascinating and compelling story lines from the discursive avalanche of official statements and private conversations available to the researcher.

The Tennessee Valley Authority (TVA) is no longer promoted as the model of how government can address problems of regional planning and economic development—as it was heavily during the periods when the book was initially published and even when the paperback edition was published in 1966. Yet Selznick's keen analytic vision of formal organizations struggling to survive as well as achieve their substantive goals, in hostile environments dominated by pre-existing organizations and deep cultural and social preconceptions, remains an inspiration to any researcher embarking on the study of a formal organization. It is little wonder that the book has been cited in over 2,000 scholarly books and articles, including more than 700 since the year 2000 alone (over half a century after it was first published), mostly by scholars in business, economics, sociology and political science.

If *TVA and the Grass Roots* needs no rediscovery, it does deserve to be discovered for the first time by at least two different bodies of readers: (1) those interested in the failures of the liberal effort to modernize and remake American government in the 20th century in both its New Deal and Great Society phases (along with those hoping for a rebirth of such efforts in the age of Obama); and (2) students of law and society interested in the social forces that shape legal doctrine and the fate of legal reform projects. When the TVA was authorized by Congress in 1933, it represented one of the most vivid examples of the New Deal's promise to reinvent government to solve the grave economic crisis of the Depression. The Tennessee Valley region (mainly Tennessee, Alabama, Mississippi and Kentucky, but including parts of Georgia,

North Carolina and Virginia as well) was an area as devastated by the Depression, and by pre-existing patterns of economic failure, as could be found in the United States. The Authority's mandate to acquire private land and assets to control flooding, and to produce and sell power, was a striking example of direct government participation in the economy—and to its critics, a large step toward socialism. Equally innovative was its emphasis on regionalism and environmental planning, themes that also struck chords of hope for change as well as resistance in a country still overwhelmingly committed ideologically to states' rights and private enterprise.

The concept of a commitment to "grass roots" (i.e., local) participation and influence came after the formal authorization of TVA, but that meme was quickly promoted by TVA's leadership in highly public ways as a rejoinder to critics. TVA was not socialism, nor to be analogized to "five year plan" type bureaucracies in the Soviet Union, precisely because it was committed to letting local actors and pre-existing organs of civil society help make choices and implement them. Against the cry (still echoing loudly in our decade) that the federal government is overriding the freedom of individuals, TVA offered a compelling case for government as an enabler of responsible individual initiative. The Authority promised to battle the environmental destruction and economic failure of much small-scale subsistence farming, not with collective farms or command-and-control regulations, but with experiments, education and practical assistance to the individual farmer, mediated by longstanding and familiar structures. The mediating structures included such grass roots institutions as the land grant universities and the agricultural extension services promoted by both parties since the middle of the previous century.

But while the sense that the TVA might be a liberal model for development both in the United States and in the emerging Third World (after World War II) was still in the ascendance when Selznick wrote up his study (and would last until well after its republication in the 1960s), Selznick could see the writing of failure on the wall. In terms that now haunt not just the New Deal, but much of the legacy of the Great Society and civil rights eras as well, *TVA and the Grass Roots* is really the story of how and why liberal reform at the national level in the U.S. tends to be subverted by locally dominant conservative forces. The substantial obstructions that the Constitution creates to sweeping change (including the bicameral structure of Congress and the non-population based allocation of seats in the Senate) means that sustained legislative support for a reform program is rare in the face of strong opposition. Selznick examined several forms of such opposition —local and federal, corporate and individual, social and political—in his prescient book.

Even with the remarkable political success of FDR and his giant Democratic majorities in Congress, ideas like TVA were always vulnerable to resistance by multiple entrenched forces with significant allies in Congress. Senator George Norris of Nebraska, who was FDR's main ally in pushing TVA through Congress, succeeded because he was an astute politician and leader of the progressive-liberal caucus in the Senate, much expanded by the 1932 election. The continued electoral strength of the Democratic Party during FDR's administrations did not assure continuous congressional backing for TVA. For the rest of the decade of the 1930s, when TVA was establishing its major policies and practices, the possibility of suffering a significant reduction in powers or even elimination at the hands of new legislation clearly haunted TVA's leaders and supporters in Congress.

As Selznick shows with brilliant clarity the Authority began from the very start a strategy of cooptation and accommodation with powerful local institutions in the states in which it operated, a strategy that would fatally compromise its ability to change the structure of inequality and economic stagnation in the region. Seeking to protect its seemingly most important progressive mission, i.e., the public ownership of land to control flooding and produce electric power, TVA's leaders sought to co-opt locally dominant and deeply conservative institutions like the state universities and the Farm Bureau by giving them influence over a substantive area of less interest, i.e., the distribution of chemical fertilizer and farming techniques to go with it. Anyone looking at the purposes and goals of TVA's commitments would have noticed little sign of mission change. Selznick, focused on the back end of methods, instruments, and techniques, could trace the way conservative interests ultimately influenced the overall mission of TVA through their success in defending and extending these technologies of power and the networks of actors and routines that went with them throughout the Authority.

The result blunted the ability of TVA to transform the social structures or conditions of the region. The ruthless domination of African Americans by whites, and of small farmers by large farmers, remained unchanged. Rather than using federal power and resources to include African Americans in the new institutions of power, and to channel investment and opportunity to small farmers, these resources and powers ended up reinforcing the hegemony of whites and large farmers. Ironically, the rhetoric of "grass roots," which might have justified cutting out existing networks of influence and creating new pathways for the region's majority of poor and excluded, was used by TVA's leadership to justify accommodating existing local power centers.

For those who think the great battles over civil rights and anti-poverty in America were won and lost in the 1960s, *TVA and the Grass Roots* is a vital political history. When we look back across the last century at the fate of progressive reform agendas in both fields, it becomes clear that the battles in the 1930s that Selznick traces here were in many respects decisive. The former states of the Confederacy represented the political bulwark of resistance to the use of government to improve the standing of African Americans and labor, including opposition to integration and unions. TVA represented a huge missed opportunity to weaken the power of entrenched elites in an important segment of southern states. When the great wave of the New Deal and the Great Society crested and the politics of reaction became more dominant nationally from the late 1960s, it was "southern" strategies that would remake the political parties and federal policy. Ironically, by then the deformation of TVA's progressive agenda from the insides of method and technique would render even its primary missions of flood control and energy production far less successful than they might have been. At the start of his remarkable rise as a national conservative leader, Ronald Reagan (working for General Electric to make the case against government ownership and regulation) used TVA as a primary rhetorical example of how government was the problem not the solution.

Selznick's innovative focus on the methods and instruments of power enabled him to see the trajectory of the New Deal reform machinery decades before it would collapse. At the time of the book's second edition in 1966, this perspective gave the work a "pessimistic" slant that Selznick goes out of the way to apologize for in his "1965 Preface" to that edition (pp. vii-x). Forty-six years later this pessimism looks like prophecy. In this regard, Selznick's approach in *TVA and the Grass Roots* bears comparison to another intellectual "pessimist" whose work has much influenced historians, sociologists and political scientists studying governance in recent years, the late French philosopher Michel Foucault. Both approached the study of power by emphasizing the importance of methods, instruments and techniques, asserting against the grain of most history that purposes, goals and objectives are almost always determined by the successes and failures of uncelebrated technologies of power. In terms that seem to anticipate Foucault's today better-known studies of power,[1] Selznick wrote in his introduction to the first edition in 1949:

> But methods are more elusive. They have a corollary and incidental status. A viable enterprise is sustained in the public eye by its goals, not its methods. Means are variable and expedient. Their history is forgotten or excused. Here again the concrete and colorful win easiest attention. Where incorrect methods leave a residue—a rubbled city or wasted countryside,—then methods may gain notice. But those means

---

[1] See Michel Foucault, *Discipline and Punish: The Birth of the Prison*, translated by Alan Sheridan (1977); Michel Foucault, *The History of Sexuality, Vol. I: An Introduction*, translated by Robert Hurley (1978); Michel Foucault, "Governmentality," translated by Rosi Braidotti and revised by Colin Gordon, in Graham Burchell, Colin Gordon and Peter Miller (eds.), *The Foucault Effect: Studies in Governmentality* (1991), pp. 87–104.

which have long-run implications for cultural values, such as democracy, are readily and extensively ignored. (P. 7.)

It is also for this focus on technologies of power that the current generation of socio-legal scholars and all of those interested in the ability of law to lead reform in democratic societies should read the *TVA and Grass Roots*. Selznick is recognized as one of the founders of the modern Law & Society movement, becoming the original director of the Center for the Study of Law & Society in 1961 and the founding chair of the Jurisprudence and Social Policy program in 1977, both at UC Berkeley. But it is Selznick's later work on jurisprudence in the 1960s and 1970s that students generally encounter. Yet his methodological strategy with its emphasis on the technologies of power offers a tremendous model for the empirical study of law. Leading contemporary work, like my colleague Lauren Edelman's path breaking studies of how businesses co-opt civil rights law,[2] reveals much the same analytic strategy in action (albeit with different theories and empirical methodologies). Other contemporary law and society scholars who follow Selznick's lead in studying law and governance through organizations and their technologies of power in this respect include Diane Vaughan, Mark Suchman, Joseph Rees, and Robert Rosen.

Philip Selznick died in June of 2010 after an extraordinarily productive career as a sociologist and legal theorist that lasted virtually to the very end.[3] It is sad that he is not here to write a preface to this new edition of his first book, but thanks to Selznick's skills as a writer and researcher those insights remain accessible to a new generation of students and concerned citizens. More than sixty years after its first publication, *TVA and the Grass Roots* remains a critical tool for either those seeking to understand the fate of the liberal legal reform projects of the 20th century or those seeking to study the emergence of new organizations and governmental institutions in the 21st. TVA was one of the first organizational fruits of the governmental revolution we know as the New Deal, launched in response to the cataclysm of the Great Depression. In the fall of 2008, the United States and indeed much of the industrialized world experienced the greatest economic cataclysm since that time. In response, a newly elected President, Barack Obama,

---

[2] See Lauren B. Edelman, "Legal Ambiguity and Symbolic Structures: Organizational Mediation of Civil Rights Law," *The American Journal of Sociology* Vol. 97, No. 6 (May, 1992), pp. 1531-1576; Lauren B. Edelman, Howard S. Erlanger and John Lande, "Internal Dispute Resolution: The Transformation of Civil Rights in the Workplace," *Law & Society Review* Vol. 27, No. 3 (1993), pp. 497-534.

[3] Philip Selznick's last book, completed just two years before his death and nearly sixty years after *TVA*, was *A Humanist Science: Values and Ideals in Social Inquiry* (Stanford University Press, 2008).

has launched a series of reform initiatives including major new laws governing health insurance and financial regulation. The fate of this new period of reform will play out over the coming years and decades. A new generation of students will have the opportunity to study the informal structures that will inevitably develop in and around the new formal organizations that the reform laws have created. In doing so they will find *TVA and the Grass Roots* an invaluable source of research strategies and tools.

JONATHAN SIMON
Adrian A. Kragen Professor of Law
University of California at Berkeley

Edinburgh
February, 2011

1965 PREFACE

IN THE SOCIAL science community *TVA and the Grass Roots* has had, I believe, several different audiences. Some readers, of course, have had a special interest in the Tennessee Valley Authority itself, as a challenging venture in public enterprise and resource development. Others have had a more general concern for the scholarly study of complex organizations. For them, the book is a case study of bureaucracy or "administrative process" and it exemplifies what might be called the "institutional approach" in organization theory. Still others have seen in the study a style of social criticism and have responded mainly to what it seemed to say about the fate of democratic aspirations in modern society. Students of political sociology tend to bridge these interests.

Since I have discussed organization theory and the institutional approach elsewhere, especially in *Leadership in Administration*, I shall confine myself here to some comments on the basic argument of the book and its intellectual background.

Viewed from one vantage point, this study stands in the tradition of Karl Marx, Gaetano Mosca, Robert Michels, Karl Mannheim, and other critics of social myths and ideologies. The TVA mythology is scrutinized in several ways. Most important is the critique of official TVA doctrine as a screen for covert, opportunistic adaptation. Here the mode of analysis used by Robert Michels in his still-remarkable *Political Parties: A Study of Oligarchical Tendencies in Modern Democracy* is especially relevant. Indeed, I had that book very much on my mind during the years just prior to 1942-43, when this research was done. Michels taught that ideals go quickly by the board when the compelling realities of organizational life are permitted to run their natural course. For Michels and his co-thinkers, words like democracy and socialism might be useful as utopian calls to action, and as cementers of solidarity, but they do not serve well as guides to policy. They do not contain the specific criteria needed to assure that the contemplated end is truly won and that the cost of winning it is acceptable. They also become readily available as protective covers behind which uncontrolled discretion can occur.

In his quest for insight into the forces that frustrate idealism, Michels did not take the easy road of criticizing the apologist for an idealized status quo; nor was he content merely to document that there is always a gap between the idea and the reality. Instead he chose to show from the history of western socialism how difficult it is for even radical idealists to avoid the tyranny of means and the impotence of ends. Means tyrannize when the commitments they build up divert us from our true objectives. Ends are impotent when they are so abstract and unspecified that they offer no principles of criticism and assessment.

Sociological realism, thus pitilessly applied to moral abstractions, can be an exercise in skepticism and an apology for passively accepting things as they are. But it is also a way of taking ideals seriously. If ideals *are* to be taken seriously, there must be genuine concern for their embodiment in action, and especially in the routines of institutional life. We cannot be content with unexamined formulae, no matter by whom expressed. We cannot forego a look behind the facade—even if it is our own.

Some years ago Alvin W. Gouldner took me to task because, as he put it; I embraced a "pathos of pessimism" and "chose to focus on those social constraints that *thwart* democratic aspirations, but neglected to consider the constraints that enable them to be *realized...*"[1] It is true that in studying the TVA I made such a choice. No doubt that choice reflected a personal concern and perhaps a personal history. But I should like to stress here that, in exploring the tendencies that inhibit democracy, I have not meant to express a generally pessimistic orientation. While I believed then, and still believe, that the corruption of ideals is easier than their fulfillment, and is in that sense more "natural," it does not follow that we should fail to treasure what is precarious or cease to strive for what is nobly conceived.

The point of anti-utopian criticism is not that it denigrates ideals. Rather it asks that such ideals as self–government be given their proper place in human affairs. Ideals are definers of aspiration. They are judgments upon us. But they are not surrogates for operative goals. The latter have the special virtue, and suffer the peculiar hardship, of striving to be reasonably adequate renderings of the moral ideal while taking due account of the human condition and the historical setting. A practical goal that does not rise to opportunities is unworthy; but one that ignores limitations invites its own corruption.

I have never supposed that the familiar apothegm, "eternal vigilance is the price of liberty" is notably pessimistic in spirit. Yet it suggests that freedom is not self-sustaining and requires special support. Our constitutional system is instinct with hope, but it is also full of sober premises about men and power. I do not think it thereby manifests a mood of tragic irony.

Nevertheless, I am not unmindful of the danger to which Professor Gouldner alluded of taking Michels too seriously. I called attention to that error myself in an essay written shortly after the publication of *TVA and the Grass Roots*.[2] There I suggested that, in considering the tyranny of means, "we should avoid looking at the devil with fascinated eyes.... Those of us who have emphasized the relevance of Michels...have done so not to damn democracy but to save it." From such critics "we learn what must be taken account of in action, for we are told how democratic organizations will develop *if* they are permitted to follow the line of least resistance."

When I said (on page 252) that an emphasis on how men are constrained and their alternatives limited "will tend to give pessimistic overtones to the analysis, since such factors as good will and intelligence will be de-emphasized," I did not mean to depreciate good will and intelligence as creative forces. I sought to remind the reader, lest he forget, that my analysis was consciously selective. I meant to warn him against just those pessimistic overtones.

I should also like to add the thought that reason, freedom, and selflessness, though they are precarious ideals, are also "determined." They are more likely to appear—they fare better—under some conditions than others. Just because this is so, we can arm ourselves against knowable threats and we can look forward with optimism to congenial conditions.

To my mind, the proper perspective of the social critic is that of the moral pragmatist. The latter does not shrink from symbolism, nor does he reject the rhetoric of hope and aspiration. He knows that a steady diet of cynicism and self-doubt can be spiritually corrosive and politically enervating. Therefore he cannot forego ideology. Yet as a pragmatist he seeks never to lose his critical sensibility, never to stop asking whether the end he has in view or the means he uses are governed by truly operative criteria of moral worth. Therefore he strives to think concretely, to look at real choices and trace their actual consequences. And the consequences he has most in mind are those that redound back on the character of the actor.

It was with the aim of looking closely and asking questions about means and ends that I approached TVA's philosophy of "grass roots" administration. I did not seek to debunk. I sought only to understand the price that is paid when ideology becomes a resource in the struggle for power. I began with the premise that in administrative life, as elsewhere, what is unscrutinized is uncontrolled and what is uncontrolled is often costly.

In this spirit the study gave most attention to two themes: (1) How were the abstractions of the "grass roots" doctrine specified operationally, that is, how did they show up in the actual course of decision-making? (2) What effect did this operative conduct have on the moral posture and competence of the TVA as a government agency?

I came to the conclusion that, whatever the theoretical validity of the administrative philosophy put forward by Chairman Lilienthal, in context and in practice it resulted in a serious weakening of the TVA's capacity to be a first-line, committed conservation agency.

I have not studied the recent history of TVA, but I note for what it is worth that on July 4, 1965 *The New York Times* carried a story headed "TVA Attacked for Strip Mining." "Long a highly regarded institution and virtually immune from criticism, TVA is coming under increasing attack from conservationists and residents of coal-producing areas for its major role in encouraging the spread of strip or surface mining for cheaper and cheaper coal." The Governor of Kentucky was quoted as saying that he believed "the people of Kentucky take a dim view of the TVA's returning such a handsome 'profit' to the Federal Treasury at the expense of ruined hillsides, poisoned streams, dead woodlands and devastated farms, a breeding ground of mosquitoes and eradicated wild life. Certainly the TVA, which is basically a conservation agency, should insist that good conservation practices be observed whenever it does business. The conscience of the Authority should not allow the destruction it today is helping to promote."

This book has to do with "the conscience of the Authority" and with the effects upon that conscience of a basic political compromise. If the current charges against the TVA have merit, it may well be that they have their roots in the story told here. For I believe that at an early period the foundations were laid for a weakness that would affect the character of the agency for a long time.

My conclusion was not merely that TVA trimmed its sails in the face of hostile pressure. More important is the fact that a right wing was built *inside* the TVA. The agricultural program of the agency was simply turned over to a group that had strong commitments, not only to a distinct ideology but to a specific constituency. This group then became a dynamic force within the TVA, able to affect programs marginal to the agricultural responsibilities of the agency but significant for conservation and rural life.

This was not a case of simple compromise made by an organization capable of retaining its internal unity. Rather, a split in the character of the agency was created. As a result, the TVA was unable to retain control over the course of the basic compromise. Concessions were demanded and won which may not have been essential if there had been fundamental unity within the organization.

If there is a practical lesson for leadership here, it is this: if you have to compromise, guard against organizational surrender. Beyond that, I hope some insight is afforded into the recurrent dilemmas of power, ideology, and organization.

PHILIP SELZNICK

Berkeley, California
August, 1965

---

[1]
"Metaphysical Pathos and the Theory of Bureaucracy," *American Political Science Review*, Vol. 49 (1955), p. 505.

[2]
"The Iron Law of Bureaucracy: Michels' Challenge to the Left," *Modern Review* (January, 1950).

PREFACE

THIS STUDY was made possible by a field fellowship granted me in 1942–1943 by the Social Science Research Council; and by the coöperation of the Tennessee Valley Authority, which opened its doors to scientific inquiry "with no strings attached." To both of these organizations I am indebted for the opportunity they created.

The materials of this inquiry were gathered during 1942–1943. The analysis was not committed to paper until three years later—a wholly incidental consequence of the war years. Subsequent developments in the TVA program and organization have not been taken into account. But since the primary interest of the study is in theoretical considerations, the delay is not, perhaps, as consequential as it might otherwise be. On the other hand, the situation within TVA as it was in 1943 represented the close of a decade of its operation, a point to be borne in mind by those interested in TVA's history for its own sake.

The files and the personnel of the Tennessee Valley Authority were the primary sources of research data. The unpublished "record" has been accorded the same status as personal interview materials, so that sources and quotations cannot always be given specific reference. I have endeavored to protect the anonymity of those in and out of the Authority who have helped me to an understanding of the TVA's methods and program. At the same time, informants on questions of detail have been restricted to those within TVA who have worked on the programs discussed. A check with the written record was made wherever possible. Interviews with officials in Washington and in the Tennessee Valley states were also of assistance.

It is hoped that a contribution has been made here toward the evolution of a theory of organization. In that sense, the study is not practical or programmatic. It is believed, however, that a practical relevance will be discerned by those involved in action who must take into account such general relations within and among organizations as are studied here. It must also be emphasized that what is presented here is only one aspect of the total TVA picture. For more general presentations of the Authority's program, the reader is referred to such volumes as David E. Lilienthal's *TVA: Democracy on the March*, C. Herman Pritchett's *The Tennessee Valley Authority: A Study in Public Administration*, and Herman Finer's *TVA: Lessons for International Application*.

It is unfortunate that the nature of the materials makes impossible explicit acknowledgment of my debt to the many individuals who gave liberally of time and faith so that I might have the materials for a realistic analysis. Without the guidance which only participants can give, much of this analysis could never have been made explicit. In addition to those unnamed, I wish to thank Daniel Bell, Edmund de S. Brunner, Patterson H. French, Max M. Kampelman, Robert S. Lynd,

Robert K. Merton, and John D. Millett. To the critical intelligence of Gertrude Jaeger there is a special obligation.

P.S.

Los Angeles
November, 1947

TVA

AND THE

GRASS ROOTS

Introduction

TVA AND DEMOCRATIC PLANNING

TVA AND DEMOCRATIC PLANNING

In this country we are very vain of our political institutions, which are singular in this, that they sprung, within the memory of living men, from the character and condition of the people, which they still express with sufficient fidelity. . .

EMERSON

WHATEVER the ultimate outcome, it is evident that modern society has already moved rather far into the age of control. It is an age marked by widening efforts to master a refractory industrial system. That a technique for control will emerge, that there is and will be planning, is hardly in question. What is more doubtful is the character and direction of the new instruments of intervention and constraint. For these have been born of social crisis, set out piecemeal as circumstances have demanded; they have not come to us as part of a broad and conscious vision. As a consequence, the foundations of a clear-cut choice between totalitarian and democratic planning have not been adequately laid; nor has the distinction been altogether clear between planning directed toward some acceptable version of the common good and planning for the effective maintenance of existing and emerging centers of privilege and power.

Democracy has to do with means, with instruments, with tools which define the relation between authority and the individual. In our time, new and inescapable tasks demand a choice among available means within the framework of increased governmental control. It is therefore especially important to examine those organizations which are proposed as contributions to the technique of democratic planning. An example of such a proposed contribution is the Tennessee Valley Authority.

On June 25, 1942, *The Times* (London) published a brief review of TVA under the heading "The Technique of Democratic Planning." *The Times* correspondent reported that he was impressed by the physical accomplishments of dam and power plant construction, but what interested him most was "the technique which the TVA had adopted with the deliberate aim of reconciling over-all planning with the values of democracy." Here *The Times* reflected what many feel to be the enduring significance of this much discussed government agency. The theme of democracy in government administration was also prominent in a widely distributed book, *TVA: Democracy on the March*, written by David E. Lilienthal, and in numerous speeches and pamphlets emanating from {*page* 4:} the Authority. In addition, much of the comment friendly to the agency has stressed its contribution to a new synthesis, one which would unite positive government—the welfare or service state—with a rigorous adherence to the principles of democracy.

What is this organization which is thought to embody an ideal so eagerly sought? What is the nature of this democratic technique? What are its implications and consequences? What will a close and critical study of the organization in action tell us about these problems? These questions have yet to be satisfactorily answered. To seek a partial answer, a study was undertaken, during 1942-1943, with attention focused primarily upon the Authority's "democratic" or "grass roots" method. This inquiry was based upon the assumption that no prior personal commitment to the TVA as a political symbol ought to interfere with a realistic examination. It was an inquiry which did not hesitate to seek out informal and unofficial sources of information. And it began with certain ideas about the nature of the administrative process which seem helpful in uncovering the underlying forces shaping leadership and policy.

The Tennessee Valley Authority was created by Congress in May, 1933, as a response to a long period of pressure for the disposition of government-owned properties at Muscle Shoals, Alabama. During the First World War, two nitrate plants and what was later known as Wilson Dam were constructed, at a cost of over $100,000,000. For the next fifteen years, final decision as to the future of these installations hung fire. The focal points of contention related to the production and distribution of fertilizer and electric power, and to the principle of government versus private ownership. Two presidential commissions and protracted congressional inquiries recorded the long debate. At last, with the advent of the Roosevelt administration in 1933, the government assumed responsibility for a general resolution of the major issues.

The TVA Act as finally approved was a major victory for those who favored the principle of government operation. The Muscle Shoals investment was to remain in public ownership, and this initial project was to be provided with new goals and to be vastly extended. A great public power project was envisioned, mobilizing the "by-product" of dams built for the purpose of flood control and navigation improvement on the Tennessee River and its tributaries. Control and operation of the nitrate properties, to be used for fertilizer production, was also authorized, although this aspect was subordinated in importance to electricity. These major powers—authority to construct dams, deepen the river {5} channel, produce and distribute electricity and fertilizer— were delegated by Congress to a corporation administered by a three-man board of directors.

If this had been all, the project would still have represented an important extension of government activity and responsibility. But what began as, and what was generally understood to be, primarily the solution of a problem of fertilizer and power emerged as an institution of far broader meaning. A new regional concept—the river basin as an integral unit—was given effect, so that a government agency was created which had a special responsibility neither national nor state-wide in scope. This offered a new dimension for the consideration of the role of government in the evolving federal system. At the same time, the very form of the agency established under the Act was a new departure. There was created a relatively autonomous public corporation free in important aspects from the normal financial and administrative controls exercised over federal organs. Further, and in one sense most important, a broad vision of regional resource development—in a word, planning—informed the conception, if not the actual powers, of the new organization.

The Message of the President requesting the TVA legislation did much to outline that perception: "It is clear," wrote Mr. Roosevelt, "that the Muscle Shoals development is but a small part of the potential public usefulness of the entire Tennessee River. Such use, if envisioned in its entirety, transcends mere power development: it enters the wide fields of flood control, soil erosion, afforestation, elimination from agricultural use of marginal lands, and distribution and diversification of industry. In short, this power development of war days leads logically to national planning for a complete river watershed involving many States and the future lives and welfare of millions. It touches and gives life to all forms of human concerns." To carry out this conception, the President recommended "legislation to create a Tennessee Valley Authority—a corporation clothed with the power of government but possessed of the flexibility and initiative of private enterprise. It should be charged with the broadest duty of planning for the proper use, conservation, and development of the natural resources of the Tennessee River drainage basin and its adjoining territory for the general social and economic welfare of the Nation."

This special regional focus and broad scope of the project have given it a character which reflects one of the major motifs of our time: the need for some sort of integral planning, especially in key problem areas. {6} It is that character which has been caught up as a model for similar projects in other areas. For the uniqueness of TVA is not that it is a government-owned power business or conservation agency, but that it was given some responsibility for the unified development of the resources of a region.

Yet it must be said that although the agency and its program have symbolized concentrated effort and planning, in fact the TVA has had little direct authority to engage in large-scale regional planning. The powers delegated to it were for the most part specific in nature, related to the primary problems of flood control, navigation, fertilizer, and power. In addition, authority to conduct studies and demonstrations of a limited nature, but directed toward general welfare objectives, was delegated to the President and by him to the Authority. This became the basis for some general surveys and demonstration work in forestry, local industrial development, community planning, and for work with coöperatives.

More important, however, is that the Act permitted such discretion in the execution of the primary purposes as would invite those in charge to recognize the social consequences of specific activities—such as the effect upon farm populations and urban communities of the creation of large reservoirs—and to assume responsibility for them. This assumption of responsibility invests the administration with an important planning function, though it is indirect and remains modifiable as circumstances may demand. In addition, there remained administrative freedom to devise methods of dealing with local people and institutions which would reflect the democratic process at work. Perhaps of equal importance is that the idea of planning associated with TVA accords this agency a central status in the consideration of the problems and the future of the Tennessee Valley region.

In the light of this weak delegation of broad planning powers, and the tendency of Congress to restrict developmental functions, it is probable that the significance of TVA in relation to democratic planning comes primarily from the infusion of specific tasks with a sense of social responsibility. In the purchase of lands, in the distribution of fertilizer and power, in personnel policy—in those functions which are a necessary part of the execution of its major and clearly delegated responsibilities—the TVA has normally taken account of the people of the area, with a view to adjusting immediate urgencies to long-term social policy. This, of course, is not the same as devising and executing a frontal plan for the reconstruction of the economy or institutions of an area. And {7} yet, whichever view is emphasized—whether one conceives of TVA's limited regional planning as a portent of fuller ventures along that line, or whether one thinks of planning as simply an adjunct of specific responsibilities—we have something to learn from a study of the organization itself and of the methods developed in the execution of its tasks.

"Organization" and "method" are key words. Wherever we turn in considering the implications of a program for democracy these terms are inevitably involved. No democratic program can be unconcerned about the objectives of a course of action, especially as they affect popular welfare. But the crucial question for democracy is not what to strive for, but by what means to strive. And the question of means is one of what to do now and what to do next—and these are basic questions in politics.

If the problem of means is vital, it is also the most readily forgotten. "Results," "achievement," and "success" are heady words. They induce submission and consent, thus summoning rewards for diligence and labor—and they also enfeeble the intellect. For the results which most readily capture the imagination are external, colorful, concrete. They are the stated goals of action. Their achievement lends reality, wholesomeness, and stature to the enterprise as a whole.

But methods are more elusive. They have a corollary and incidental status. A viable enterprise is sustained in the public eye by its goals, not its methods. Means are variable and expedient. Their history is forgotten or excused. Here again the concrete and colorful win easiest attention. Where incorrect methods leave a visible residue—a rubbled city or wasted countryside,—then methods may gain notice. But those means which have long-run implications for cultural values, such as democracy, are readily and extensively ignored.

When we speak of methods, we speak in the same breath of instruments. Policies, decisions as to "how to proceed," require execution. Execution in turn implies a technology. We are familiar with the kind of technology which includes machines and tools of all sorts, handled and manipulated in more or less obvious ways. We are even reasonably familiar with the technology of economic and military organization, geared to the achievement of technical objectives, qualified and informed by the criteria of efficiency. But when we move into that area of technology which is related to the creation, defense, or reintegration of values, such as democracy, we find ourselves less assured. Yet the significance of this noneconomic technology, under the conditions of mass society and cultural disintegration, is of primary importance for whatever {8} we may wish to do about that vague but demanding reality which we call our "way of life." Propaganda agencies, mass parties, unions, educational systems, churches, and governmental structures have a common aspect in that, more or less directly, they work upon and seriously affect the evolving values, the spirit, of contemporary society. Furthermore, there is a growing tendency for this effect to be conscious, to become an ordered technology available to those who have a stake in changing sentiment or social policy.

One of the pervasive obstacles to the understanding and even the inspection of this technology is ideology or official doctrine. By the very nature of their function, all those forces which are concerned about the evolution of value-impregnated methods, or public opinion itself, have a formal program, a set of ideas for public consumption. These ideas provide a view of the stated goals of the various organizations—political or industrial democracy, or decentralization, or the like—as well as of the methods which are deemed crucial for the achievement of these goals. It is naturally considered desirable for the attention of observers to be directed toward these avowed ideas, so that they may receive a view of the enterprise consistent with the conception of its leadership. All this in the often sincere conviction that precisely this view is in accord with the realities of the situation and best conveys the meaning and significance of the project under inspection.

However much we may be impressed by what a group says about its methods or its work, there is adequate justification for uneasiness and doubt. This doubt has its source in our general understanding of the persistent tendency for words to outrun deeds, for official statement and doctrine to raise a halo over the events and activities themselves. That this is a natural disposition among responsible men is well understood, and a gap of some sort between the idea and the act is normally expected. But what is less well understood, or at least less generally applied to objects of public esteem, is the tendency for ideas to reflect something more than enthusiasm or more or less pardonable pride. The functions of a doctrine may be more subtle and more significant, related to the urgent needs of leadership and to the security of the organization itself. Such functions, when relevant, cast a deeper shadow and indicate the need for more searching questions. In Part I of this study, we have critically analyzed TVA's official doctrine in relation to democratic planning—the policy of grass-roots administration as a contribution to democracy. The analysis points to underlying issues and problems not directly evident when we speak of the normal and anticipated gap between avowed statement and actual practice. {9}

Though official statements and theories are important, an undue concentration upon what men say diverts attention from what they do. This is especially true with respect to the methods utilized in the execution of a program, for these are particularly difficult to view realistically. It is often sufficiently troublesome to attain a clear picture of the formal, stated methods in use, without pressing inquiry as to the less obvious but vital informal behavior of key participants. Yet it is precisely into the realm of actual behavior and its significance for evolving structures and values that we must move if this kind of inquiry is to realize its possibilities.

The instruments of planning are vitally relevant to the nature of the democratic process. The TVA is many things, but most significant for our purposes is its status as a social instrument. It is this role as instrument with which this study is directly concerned. Or, to emphasize another word, it is TVA as an organization to which our attention is directed. Thus it is not dams or reservoirs or power houses or fertilizer as such, but the nature of the Authority as an ordered group of working individuals, as a living institution, which is under scrutiny.

In searching out organizational behavior and problems as keys to understanding the implications of TVA for democratic planning, we are entering a field of inquiry which probes at the heart of the democratic dilemma. If democracy as a method of social action has any single problem, it is that of enforcing the responsibility of leadership or bureaucracy. A faith in majorities does not eliminate the necessity for governance by individuals and small groups. Wherever there is organization, whether formally democratic or not, there is a split between the leader and the led, between the agent and the initiator. The phenomenon of abdication to bureaucratic directorates in corporations, in trade unions, in parties, and in coöperatives is so widespread that it indicates a fundamental weakness of democracy. For this trend has the consequence of thrusting issues theoretically decided by a polity into the field of bureaucratic decision.

The term "bureaucracy" has an invidious connotation, signifying arbitrary power, impersonality, red tape. But if we recognize that all administrative officials are bureaucrats, the bishop no less than the tax collector, then we may be able to understand the general nature of the problem, separating it from the personal qualities or motives of the individuals involved. Officials, like other individuals, must take heed of the conditions of their existence. Those conditions are, for officials, organizational: in attempting to exercise some control over their own {10} work and future they are offered the opportunity of manipulating personnel, funds, and symbols. Among the many varied consequences of this manipulation, the phenomena of inefficiency and arbitrariness are ultimately among the least significant. The difference between officials and ordinary members of an organized group is that the former have a special access to and power over the machinery of the organization; while those outside the bureaucratic ranks lack that access and power.

If we are to comprehend these bureaucratic machines, which must play an indispensable role in any planning venture, it is essential to think of an organization as a dynamic conditioning field which effectively shapes the behavior of those who are attempting to remain at the helm. We can best understand the behavior of officials when we are able to trace that behavior to the needs and structure of the organization as a living social institution.

The important point about organizations is that, though they are tools, each nevertheless has a life of its own. Though formally subordinated to some outside authority, they universally resist complete control. The use of organizational instrumentalities is always to some degree precarious, for it is virtually impossible to enforce automatic response to the desires or commands of those who must employ them. This general recalcitrance is recognized by all who participate in the organizational process. It is this recalcitrance, with its corollary instability, which is in large measure responsible for the enormous amount of continuous attention which organizational machinery requires. There are good reasons, readily grasped, for this phenomenon.

The internal life of any organization tends to become, but never achieves, a closed system. There are certain needs generated by organization itself which command the attention and energies of leading participants. The moment an organization is begun, problems arise from the need for some continuity of policy and leadership, for a homogeneous outlook, for the achievement of continuous consent and participation on the part of the ranks. These and other needs create an intricate system of relationships and activities, formal and informal, which have primarily an internal relevance. Thus leadership is necessarily turned in upon itself. But at the same time, no organization subsists in a vacuum. Large or small, it must pay some heed to the consequences of its own activities (and even existence) for other groups and forces in the community. These forces will insist upon an accounting, and may in self-defense demand a share in the determination of policy. Because of this outside pressure, from many varied sources, the attention of any {11} bureaucracy must be turned outward, in defending the organization against possible encroachment or attack.

These general considerations, which have been stated here in a summary way, should lead to a more discerning study of any administrative agency. They direct us (1) to seek the underlying implications of the official doctrine of the agency, if it has one; (2) to avoid restriction to the formal structure of the organization, as that may be outlined in statutes, administrative directives, and organization charts; and (3) to observe the interaction of the agency with other institutions in its area of operation. Throughout, a search for the internally relevant in organizational behavior, especially that which is related to self-defensive needs, is a primary tool of such analysis.[1]

It will probably bear emphasis that the significance of TVA for democratic planning lies not so much in its program, or in its accomplishments, as in its methods and in its nature as an organization. Even though its planning powers are limited, the TVA does represent an experiment, an adventure in executing broad social responsibilities for the development of a unified area. Furthermore, its type of organization is proffered as a model for governmental planning in other areas. This point has been clearly recognized within TVA itself:

> Few of the activities of TVA are unique as public responsibilities. The Government of the United States has been constructing waterways and building works for flood control for more than a century. State and Federal agencies have engaged in technical research, and surveys of mineral and forestry resources have been carried on with public funds for many years. The TVA is not the first instance in which the Federal Government has sold electric power. Aid to and stimulation of business opportunities in industrial development, employment, farming, and other fields has become a familiar role of Government, State and National. {12}
>
> It is in the integration and the correlation on a regional basis of these various activities under a single, unified management that the Tennessee Valley Authority represents a pioneer undertaking of government. For the first time a President and Congress created an agency which was directed to view the problems of a region as a whole.[2]

If the power granted to the Authority was not sufficient fully to execute that broad responsibility, still the vision has remained. It is the conception of an administrative instrument created to fulfill necessary planning functions within the framework of democratic values.

If TVA as instrument is the focus of attention, and if we are prepared to think of the Authority as a living social organization, we may expect that in one way or another, the Authority will have been caught up in and shaped by its institutional environment. This expectation becomes especially relevant as we note (1) the TVA's official avowal of a special democratic relation to certain local institutions "close to the people," a doctrine which will be discussed in detail below; and (2) that TVA did not arise out of the expressed desires of the local area, and consequently was faced with a special problem of adjustment. Each of these points lends weight to the anticipation that in the Authority's relation to its own grass roots we may find significant material of general interest to those who wish to learn the lessons of the TVA experience.

Given such an anticipation, the problem for this inquiry became one of finding a significant vantage point from which to examine this grass-roots relationship. The question thus posed required some sort of theory, a set of ideas which could point a way to the most vital aspects of the situation. The theory which seemed to make sense in the light of a general understanding of the materials was so formulated as to bring together in a single over-all analysis (1) the avowed contribution of TVA to democratic planning, through a grass-roots method of executing its responsibilities; (2) the self-defensive behavior of the organization as it faced the need to adjust itself to the institutions of its area of operation; (3) the consequences for policy and action which must follow upon any attempt to adjust an organization to local centers of interest and power. Put in a few words, this involved the hypothesis that the Authority's grass-roots policy as doctrine and as action must be understood as related to the need of the organization to come to terms with certain local and national interests; and that in actual practice this procedure resulted in commitments which had restrictive consequences for the policy and behavior of the Authority itself. {13}

In order to handle this problem most effectively, it has been found necessary to introduce a concept which, while not new, is somewhat unfamiliar. This is the idea of *coöptation*[3]—often the realistic core of avowedly democratic procedures. To risk a definition: *coöptation is the process of absorbing new elements into the leadership or policy-determining structure of an organization as a means of averting threats to its stability or existence.* With the help of this concept, we are enabled more closely and more rigorously to specify the relation between TVA and some important local institutions and thus uncover an important aspect of the real meaning and significance of the Authority's grass-roots policy. At the same time, it is clear that the idea of coöptation plunges us into the field of bureaucratic behavior as that is related to such democratic ideals as "local participation."

Coöptation tells us something about the process by which an institutional environment impinges itself upon an organization and effects changes in its leadership, structure, or policy. Coöptation may be formal or informal, depending upon the specific problem to be solved.

Formal coöptation.—When there is a need for the organization to publicly absorb new elements, we shall speak of formal coöptation. This involves the establishment of openly avowed and formally ordered relationships. Appointments to official posts are made, contracts are signed, new organizations are established—all signifying participation in the process of decision and administration. There are two general conditions which lead an organization to resort to formal coöptation, though they are closely related:

1. When the legitimacy of the authority of a governing group or agency is called into question. Every group or organization which attempts to exercise control must also attempt to win the consent of the governed. Coercion may be utilized at strategic points, but it is not effective as an enduring instrument. One means of winning consent is to coöpt into the leadership or organization elements which in some way reflect the sentiment or possess the confidence of the relevant public or mass and which will lend respectability or legitimacy to the organs of control and thus reëstablish the stability of formal authority. This device is widely used, and in many different contexts. It is met in colonial countries, where the organs of alien control reaffirm their legitimacy by coöpting native leaders into the colonial administration. We find it in {14} the phenomenon of "crisis-patriotism" wherein normally disfranchised groups are temporarily given representation in the councils of government in order to win their solidarity in a time of national stress. Coöptation has been considered by the United States Army in its study of proposals to give enlisted personnel representation in the courts-martial machinery—a clearly adaptive response to stresses made explicit during World War II. The "unity" parties of totalitarian states are another form of coöptation; company unions or some employee representation plans in industry are still another. In each of these examples, the response of formal authority (private or public, in a large organization or a small one) is an attempt to correct a state of imbalance by formal measures. It will be noted, moreover, that what is shared is the responsibility for power rather than power itself.

2. When the need to invite participation is essentially administrative, that is, when the requirements of ordering the activities of a large organization or state make it advisable to establish the forms of self-government. The problem here is not one of decentralizing decision but rather of establishing orderly and reliable mechanisms for reaching a client public or citizenry. This is the "constructive" function of trade unions in great industries where the unions become effective instruments for the elimination of absenteeism or the attainment of other efficiency objectives. This is the function of self-government committees in housing projects or concentration camps, as they become reliable channels for the transmission of managerial directives. Usually, such devices also function to share responsibility and thus to bolster the legitimacy of established authority. Thus any given act of formal coöptation will tend to fulfill both the political function of defending legitimacy and the administrative function of establishing reliable channels for communication and direction.

In general, the use of formal coöptation by a leadership does not envision the transfer of actual power. The forms of participation are emphasized but action is channeled so as to fulfill the administrative functions while preserving the locus of significant decision in the hands of the initiating group. The concept of formal coöptation will be utilized primarily in the analysis of TVA's relation to the voluntary associations established to gain local participation in the administration of the Authority's programs.

Informal coöptation.—Coöptation may be, however, a response to the pressure of specific centers of power within the community. This is not primarily a matter of the sense of legitimacy or of a general and diffuse {15} lack of confidence. Legitimacy and confidence may be well established with relation to the general public, yet organized forces which are able to threaten the formal authority may effectively shape its structure and policy. The organization faced with its institutional environment, or the leadership faced with its ranks, must take into account these outside elements. They may be brought into the leadership or policy-determining structure, may be given a place as a recognition of and concession to the resources they can independently command. The representation of interests through administrative constituencies is a typical example of this process. Or, within an organization, individuals upon whom the group is dependent for funds or other resources may insist upon and receive a share in the determination of policy. This type of coöptation is typically expressed in informal terms, for the problem is not one of responding to a state of imbalance with respect to the "people as a whole" but rather one of meeting the pressure of specific individuals or interest groups which are in a position to enforce demands. The latter are interested in the substance of power and not necessarily in its forms. Moreover, an open acknowledgment of capitulation to specific interests may itself undermine the sense of legitimacy of the formal authority within the community. Consequently, there is a positive pressure to refrain from explicit recognition of the relationship established. This concept will be utilized in analyzing the underlying meaning of certain formal methods of coöperation initiated in line with the TVA's grass-roots policy.

Coöptation reflects a state of tension between formal authority and social power. This authority is always embodied in a particular structure and leadership, but social power itself has to do with subjective and objective factors which control the loyalties and potential manipulability of the community. Where the formal authority or leadership reflects real social power, its stability is assured. On the other hand, when it becomes divorced from the sources of social power its continued existence is threatened. This threat may arise from the sheer alienation of sentiment or because other leaderships control the sources of social power. Where a leadership has been accustomed to the assumption that its constituents respond to it as individuals, there may be a rude awakening when organization of those constituents creates nucleuses of strength which are able to effectively demand a sharing of power.

The significance of coöptation for organizational analysis is not simply that there is a change in or a broadening of leadership, and that this is an adaptive response, but also *that this change is consequential for the character and role of the organization or governing body*. Coöptation {16} results in some constriction of the field of choice available to the organization or leadership in question. The character of the coöpted elements will necessarily shape the modes of action available to the group which has won adaptation at the price of commitment to outside elements. In other words, if it is true that the TVA has, whether as a defensive or as an idealistic measure, absorbed local elements into its policy-determining structure, we should expect to find that this process has had an effect upon the evolving character of the Authority itself. From the viewpoint of the initiators of the project, and of its public supporters, the force and direction of this effect may be completely unanticipated.

The important consideration is that the TVA's choice of methods could not be expected to be free of the normal dilemmas of action. If the sentiment of the people (or its organized expression) is conservative, democratic forms may require a blunting of social purpose. A perception of the details of this tendency is all important for the attempt to bind together planning and democracy. Planning is always positive—for the fulfillment of some program,—but democracy may negate its execution. This dilemma requires an understanding of the possible unanticipated consequences which may ensue when positive social policy is coupled with a commitment to democratic procedure. The description and analysis which follows, in tracing the consequences of TVA's grass-roots policy for the role and character of the organization, may cast some light upon that problem.[4]

---

[1] It appears that this institutional approach (not, of course, original with the author) to the study of administrative organization may be the avenue to an enlargement of the horizon of inquiry in this field. In a sense, this approach and this study are a response to such criticism as that voiced by Donald Morrison in his review of the series, *Case Reports in Public Administration:* "To put the matter succinctly, the subject-matter of public administration has been defined so as to leave a no-man's land of significant problems, flanked on one side by the students of administration and on the other by political theorists. The problems thus isolated have their origin in the fact that in its fundamental aspect administration is governance.... One such problem, perhaps the most urgent, is to develop and strengthen ways of insuring that government by the bureaucracy does

not destroy the democratic pattern of our society. Unless it is assumed that such insurance lies in the perfection of organizational structure and techniques of fiscal and personnel management, the present series of case studies does not deal with this matter. Many persons believe that the TVA experiment is suggestive of ways of democratizing bureaucratic government. Ten TVA studies are published in *Case Reports,* but none deals with the integration of the TVA program into the social and economic life of the area" (*Public Administration Review,* V, l [Winter, 1945], 85).

[2] "The Widening of Economic Opportunity through TVA," pamphlet adapted from an address by David E. Lilienthal, Director, TVA, at Columbia University, New York, N.Y., January, 1940 (Washington: Government Printing Office, 1940), p. 15.

[3] With some modifications, the following statement of the concept of coöptation is a repetition of that presented in the author's "Foundations of the Theory of Organization," *American Sociological Review,* XIII, 1 (February, 1948), pp. 33–35. For further discussion of coöptation see below, pp. 259–261.

[4] The notion of "unanticipated consequence" referred to in this section is central to this study. See below, pp. 253–259, for a theoretical statement of the problem.

Part One

OFFICIAL DOCTRINE

CHAPTER I

THE IDEA OF A "GRASS-ROOTS" ADMINISTRATION

"Administration" means more than organization charts. Go in one administrative direction and you have loss of liberty; go far enough and you have decisions enforced by the Gestapo and the lash. Go in the other direction and you have people participating in the decisions of their government actively and with considerable zeal, an increase in freedom and the corresponding increase in responsibility and discipline. Proceeding on this basis, I began a series of public statements on organizational characteristics of the TVA, particularly decentralization of administration as a method of securing the participation of the people of the Valley in the TVA undertaking.[1]

DAVID E. LILIENTHAL

THE TENNESSEE VALLEY AUTHORITY has been the subject of widespread comment and study in all parts of the world. In Central Europe, in the Philippines, in Palestine, in China, wherever, indeed, new methods of approach to the problems of resource development and social planning have been discussed, the TVA idea has been in the forefront. TVA has become not merely an administrative model and prototype, but a symbol of the positive, benevolent intervention of government for the general welfare.

In America, too, the TVA is unquestionably a rallying-point for those who favor a welfare state. The defense of the TVA and the extension of its methods are accepted among these groups as an elementary duty. What is known in the United States as the progressive movement (essentially the forces which comprised the popular base of the New Deal) has treated the TVA as a symbol of its aspirations and has been quick to muster its forces behind the agency and the ideas it symbolizes whenever occasion demanded. This has been evident in the unequivocal defense of the TVA organization itself in controversies over finances and accountability; and especially in the vigorous espousal of the TVA idea as a model for regional development in other areas of the United States. For these forces, support of the TVA is a ready criterion of political acceptability.

It is primarily as a symbol that TVA excites allegiances and denunciation. In its capacity as symbol, the organization derives meaning {20} and significance from the interpretations which others place upon it. The halo thus eagerly proffered is in large measure a reflection of the needs and problems of the larger groups which require the symbol and use it. Hence controversy over the TVA may proceed irrespective of a close examination of the organization itself. When examinations are undertaken, they are made to exalt TVA or condemn it: on the one hand, as positive government liberating and advancing the people whom it serves, or, on the other, as a nefarious encroachment on free enterprise, with subversive intent. This is not surprising, for the symbol is necessarily caught up in and manipulated by the broader issues and forces involved in social conflict. Since it is in the context of the larger struggle that basic political decisions are made, it is to be expected that the TVA should there lose its special identity and become merely an instrument and a focal point of attack and defense.

Yet however significant its symbolic feature may be, TVA is and must be more than an idea. It is a living organization in a concrete social environment. From that obvious statement many things follow. The special leadership of such an organization is divided from the general leadership of its diffuse support by the quality of responsibility. To those for whom the idea is primary, the symbol is enough and the actual organization which embodies it at any particular time somewhat irrelevant. But the symbol become apparatus generates its own problems for whose solution the administrative leadership is directly responsible. These new problems are technical, and go beyond political loyalty to an objective; they arise out of the need to weigh means and judge consequences in the context of practical action. The administrative leadership is turned in upon itself, preoccupied with the tools at hand and with the concrete choices which must be made to implement the general policy it seeks to execute. Whatever the ultimate national implications of its choices, the organization in action must deal primarily with those who are immediately and directly affected by its intervention. In order to endure, it cannot depend only or even primarily upon the diffuse support of elements not directly involved in its work; its administrative leadership must find support among local institutions, and develop smooth working relationships with them. It must avoid a continuous atmosphere of crisis and conflict which may lead in the first instance to disorganization and frustration and in the long run to the upcropping of significant threats to the very existence of the organization itself. In short, the institution must seek some sort of equilibrium with the environment in which it lives. {21}

To point to such an adjustment, or the need for it, is to speak of the normal course of events. Experienced participants in the organizational process are familiar with the continuous striving for adjustment, though few have spoken self-consciously of it. The management of its details is accepted as part of the ordinary common sense of administrative leadership. But it is often observable that where institutions have symbolic meaning beyond (and often irrelevant to) their own structure and behavior, a blindness to the less rosy aspects of organizational life arises. Explicit statements which trace the history of the organization in terms of compromise and mediation are rejected out of hand, or at best, when accepted, are deemed shocking exposures.

If we understand that the leaders of an operating organization have problems separate from those who function primarily in the realm of symbols, then it should be of particular interest to examine the ideas and theories of that administrative leadership. Such an examination may reveal meanings evolving from the problems of the organization as such, and bearing implications not altogether consonant with the total view of those who accept and support the symbol-at-large.

The TVA is a particularly valuable subject for study in these terms. For its leaders have been especially active—more so than most other governmental agencies, though there are few which do not venture at all upon this road—in propagating a systematic formulation of its own meaning and significance. This self-analysis of the role of TVA has been elaborated at length and presented to the public at every possible opportunity. Far from being a casual or incidental analysis, it is presented as a fundamental key to an understanding of the Authority. It has been made a part of almost all important speeches and publications which describe the agency to its special publics and to the world. Inside the organization, strenuous efforts are made to indoctrinate all who are involved in administration on a policy-affecting level. The elaboration of this interpretation has been, indeed, a well-developed and effective exercise in administrative self-consciousness.

This chapter will be devoted to a statement of that doctrine. What follows, therefore, represents a paraphrase of official TVA doctrine, as that has been formulated verbally and in various documents, some published, others not. An attempt is made to bring together most of the arguments presented by TVA officials in support of the grass-roots policy, and thus to present a rounded formulation of the TVA viewpoint. This seems especially desirable because of the critical analysis which is detailed in succeeding chapters. Hence, although responsibility {22} for the accuracy with which TVA views have been reflected in this chapter must be accepted by the author, it should be understood that the intention is to present the ideas of the TVA administration and not his own.

THE NEED FOR DECENTRALIZATION OF ADMINISTRATION

David E. Lilienthal, until 1946 Chairman of the Board of Directors of the TVA, has taken the leading role in public exposition of the official TVA doctrine of grass-roots administration. While presented by him partly in personal terms, the idea is nevertheless organically related to the Authority as a whole. Lilienthal has specifically stated that the fundamentals of the theory must be credited to his colleague on the Board, Dr. Harcourt A. Morgan, former President of the University of Tennessee. Moreover, within the organization, at in-service training conferences, administrative seminars, and in numerous memorandums on tangential subjects, the theory is repeatedly discussed as basic organization policy. Its significance therefore extends far beyond the individual views of Lilienthal, though he may have provided a personal coloration for bare essentials which might otherwise be differently, and perhaps less forcefully or sympathetically, presented.

The context of the TVA thesis is the world-wide trend toward centralization and, concomitantly, the growing responsibilities of national government. Centralization has been proceeding apace in all fields of human organization. Efficiency has been, in this view, a rigorous leveler, erasing the diversity of individual enterprise and local control in the interests of large hierarchized units. This process operates not in government alone but in many other fields as well, and always brings with it like and ambiguous consequences. In exchange for the benefits of order and coördination, initiative has been stifled and the power of decision — indispensable element of democratic action — lodged in far-off places, remote from the beneficial influences of local areas which become merely the objects of bureaucratic manipulation. Small businessmen, the independent artisan, and farmer alike, have felt the enervating effect of the concentration of economic and social control. By a similar logic, small nations, too, have been unable to endure alone, and have reluctantly found their way into some broader hegemony which provides security in exchange for liberty.

In the wake of the general centralization of social (and especially economic) life, has followed inevitably the centralization of public authority. But this, like the centralization in other fields, is not unequivocally {23} bad. Just as centrally managed private enterprises have achieved lowered costs, more efficient and wider distribution, and the advancement of science and invention, so too, centralized government has brought improvements which cannot be denied out of hand. Above all, it is necessary to see that in the ambiguous results of centralization there is a problem, a dilemma, which must be recognized explicitly and boldly faced so that new techniques of organization may be devised which, while preserving the essential good, will eliminate the more critical evils. This problem is pressing, for the disappearance of small units and local controls "lays bare the peculiar hazard of this modern world: the danger implicit in vast size, the disaster consequent when power is exercised far from those who feel the effect of that power, remote and alien to their lives."[2]

The recent history of American democracy has been in significant part, according to TVA doctrine, a history of the simultaneous broadening of the responsibilities and the field of intervention of national government. Although not without difficulties, and certainly without complete uniformity, the executive, legislative, and judicial departments of the federal government have alike accepted a widened view of the fields of regulation and positive construction in which Washington agencies might operate. These new national responsibilities have been accepted in response to (1) acute needs, such as unemployment, impossible to ignore and yet beyond the power of the states to handle; (2) the demands of large interest groups, including labor and agriculture, organized on a national scale and viewing the power of the federal government as both objective and instrument; (3) the growth of centralized industry requiring the counterbalance of a federal government strong enough to meet it on something like equal terms; and (4) the growth of collectivist ideology, supporting and justifying the trend toward over-all integration. National problems have demanded national recognition, and doubtless will continue to do so. This trend is irreversible; it would be idle to attempt to turn back the clock, to return to methods which offer no answer to the real and urgent problems posed by modern society and its technology.

At the same time, we are told, the question must be raised: is it necessary that the exercise of federal functions be identified with top-heavy organizations centered in and administered from Washington? For the most part this identification has been made. As statutes have {24} been enacted recognizing new obligations of the national government, their administration has been delegated to existing departments and *ad hoc* agencies with headquarters in Washington, thus increasing the staffs, responsibilities, and functions centered at the national capital. If this procedure has sometimes effectively provided the services required, it has done so while retaining and extending the basic conception of a national government centrally administered.

This historic extension of centralized power has, in Lilienthal's view, created a justified feeling of uneasiness and distrust. Ordinary people and, increasingly, men in responsible posts in business and government have been understandably fearful of an unchecked growth of a vast administrative apparatus in Washington. It is in this proliferation of Washington-oriented agencies rather than in the mere grant of power to the federal authority that there is reason for fear. "This country is too big for such a pyramiding of direct responsibilities. For in spite of our triumphs over time and space, Washington is still remote from the average citizen and is sheltered from participation in his daily struggles."[3]

An excessively centralized government is inherently disqualified, at least in the United States, from fully promoting the welfare of its citizens. This is a consequence of the significant differences in attitude and custom which are the elements of diversity within our general cultural and political unity. These differences cannot adequately be safeguarded when programs are formulated purely in national terms. Methods of approach to local people in such administrative contexts as employment or land purchase may well be homogeneous if we consider only the interests or convenience of the bureaucracy. But the people treasure their special folkways, and protect them as basic elements of welfare and security.

This fundamental evil of overcentralized government is accompanied by another: the inhibition of action through the proliferation of red tape. A centralized agency, remote from the field of operation, lays a deadening hand upon its officers "on the line" by relieving them of the responsibility for significant decision. In the interests of standardized practice and accountability in detail, the national headquarters is driven to insist that all important problems, and many that are not so important, be referred up the hierarchy, "through channels," for action. The {25} consequences of this procedure are well known: delays when action is contemplated and the stifling of initiative in the face of a justified fear of being pitted against the resistance of the bureaucracy.

From these observations Lilienthal derives a fundamental dilemma. On the one hand, it must be conceded that increasingly large powers ought to be intrusted to the federal government, for there are too many basic problems which cannot be handled through the organs of local control; on the other hand, the centralization of large powers is always a menace to democracy. "It must be recognized that there is genuine peril if the powers of federal government are hopelessly outdistanced by the trend to centralized control in industry and commerce and finance"; still, "the dangers of centralized administration are all too evident. They cannot be ignored."[4]

For his answer to this dilemma, Lilienthal has turned to (and widely publicized) the formulations of an earlier commentator on American democracy, Alexis de Tocqueville. De Tocqueville outlined a distinction between centralized government and centralized administration, a distinction which has served as the theoretical anchor of the grass-roots answer to the dilemma of centralization.

> Centralization is a word in general and daily use, without any precise meaning being attached to it. Nevertheless, there exist two distinct kinds of centralization, which it is necessary to discriminate with accuracy.
>
> Certain interests are common to all parts of a nation, such as the enactment of its general laws and the maintenance of its foreign relations. Other interests are peculiar to certain parts of the nation, such, for instance, as the business of the several townships. When the power that directs the former or general interests is concentrated in one place or in the same persons, it constitutes a centralized government. To concentrate, in like manner in one place the direction of the latter or local interests, constitutes what may be termed a centralized administration.[5]

Lilienthal has endorsed and often quoted the following statement of De Tocqueville:

> Indeed, I cannot conceive that a nation can live and prosper without a powerful centralization of government. But I am of the opinion that a centralized administration is fit only to enervate the nations in which it exists, by incessantly diminishing their local spirit. Although such an administration can bring together at a given moment, on a given point, all the disposable resources of a people, it injures the renewal of those resources. It may ensure a victory in the hour of strife, but it gradually

24

relaxes the sinews of strength. It may help admirably the transient greatness of a man, but not the durable prosperity of a nation.[6] {26}

De Tocqueville's strictures set a serious and enduring problem, but of course they provide no solution. Yet the distinction between centralized government and centralized administration points the way, in Lilienthal's view, to a solution in modern terms. It is a way which appears to offer an alternative to the acceptance of "topheavy, cumbersome, centralized administration as the price of granting necessary powers to our federal government." The answer lies in a new solution to the way in which the powers of the national government are administered.

There is, of course, no single remedy, but it is believed in TVA that the theory of administrative decentralization, properly understood and wisely applied, will serve to check the growing tendency toward excessive centralization of federal administration. This is not a new idea, for there has been for some time an increasing awareness in Washington of the need for a greater measure of decentralization. But for the most part we have seen no fundamentally new departure, for the attempt by federal departments to set up regional offices and otherwise establish important nucleuses of administration in the field has left the old structure of control intact. The reins are held by the central offices and their various control divisions; these in turn are subject to the manifold restrictions (such as laws regulating procurement and accountability) which serve to crystallize and perpetuate the existing system of centralized control.

There has been, however, one federal agency which serves as an example of and an experiment in the decentralization of federal functions. This is the Tennessee Valley Authority, "the boldest and perhaps most far-reaching effort of our times to decentralize the administration of federal functions. If it succeeds, if its methods prove to be sound, we shall have added strength to the administrative defenses which protect the future of our beleaguered democracy."[7] The TVA is invested with the authority of the national government, and derives its power from the exercise by the federal government of its constitutional prerogatives as interpreted in our time. At the same time, administration of these powers is effectively decentralized.

When the TVA was established, the programmatic responsibilities of the federal government for the control of the natural resources of water and land were brought together and treated as a unit in order to deal effectively with the watershed of the Tennessee River as an integrated area, a single problem in resource development. Thus one agency, located in the area of operation, represented the federal government in {27} relation to a complex local problem. The TVA was not appended to any existing department, but was given freedom of action and a substantial measure of autonomy, subject only to the direct control of the President and Congress. The TVA Act, the agency's basic charter, was so framed as to give the Authority the power to make its own decisions, and thereby such flexibility as would make possible a maximum adjustment to local conditions. This delegation of discretion is the heart of the idea of grass-roots administration.

The broad significance which Lilienthal attributes to the TVA as a new departure in public administration is made additionally clear by his response to the challenge of James Burnham's widely noted book, *The Managerial Revolution*. This book singled out the TVA as an example of the social revolution, now encompassing the earth, which, as Burnham claimed, is raising the managers, among whom are the administrators of great corporations, to the heights of social power. It asserted further that this new class would in its turn become an ex-ploiting group, using specialized skill and social position for its own class benefit. Lilienthal agrees that the extension of public enterprise has been great, and that "government which has undertaken a major responsibility for production will of course depend increasingly upon men in public service skilled in the art and science of management."[8] But he denies that this is inevitably a threat to democracy. We can, he asserts, introduce counter-measures which will offset the managerial trend. Among these measures is administrative decentralization:

> That we are in grave danger of putting these fetters upon our arms no one can deny. But we do have a choice. We can choose, if we will, between an exploiting managerial class and managers bound by the principles of public service and democratic methods. Indeed managerial technicians, clothed as they must be with the great power of their skill and of society's reliance upon them, ought to assume leadership in protecting American democratic society from exploitation by the managers themselves. That leadership can be expressed, to take one example, by a constant search by public administrators for methods of avoiding overcentralization in administration. Every sophisticated manager knows that tyranny and exploitation feed upon excessive centralization of administration. He knows that overcentralized admin-istration dries up the well-springs of initiative, of energy, and of independence in any organization and most of all in government institutions. It is overcentralization that gives a "clique of headquarters courtiers" an opportunity to maneuver and flatter their way to power which they are not qualified by their abilities to exercise. Absentee government is the quickest way to raise up the exploiting managerial class that Mr. Burnham's book predicts with such confidence. But these prophecies need {28} not be fulfilled; we do have a choice, for the hazards of managerial exploitation can be diminished by skillful efforts in the direction of decentralized administration of centralized auth-ority.[9]

Thus the social problem to which the grass-roots theory is direct-ed is of sweeping proportions. And though the technique of federal de-centralization is not presented monolithically, as an only answer or a panacea, still it is clear that, in the mind of its chief proponent, it is more than a modest device linked to special problems within the federal system.

MINIMUM ESSENTIALS

In a decade of existence as a decentralized federal agency, the TVA had the opportunity to formulate the special requirements which differentiate actual decentralization from the kind which represents little more than lip service. The goals have been variously stated, but uniformly include the following elements:

1. *The responsible agency in the area of operation is permitted the freedom to make significant decisions on its own account.* This assumes that the organization is to be independent, that it will not be made a part of some larger administrative agency having the power to write rules and regulations signed by the top administrator in Washington and hence immediately applicable with the force of law to the local operating organization. It further assumes that the field officers will be of a capacity and standing which will permit them to exercise broad discretion in adapting general policies to particular local situations, and to do this in such a way that their authority in local matters will be recognized by the local public. Moreover, this independent agency, if it engages in business functions, is to be permitted wide corporate freedom in the exercise of such managerial responsibilities as the selection of personnel, procurement, and disposition of operating funds.

2. *There must be active participation by the people themselves in the programs of the public enterprise.* The services of state and local agencies are to be utilized, with the federal government providing leadership that will strengthen rather than weaken or eliminate the existing agencies. Management should devise means to enlist the active and conscious participation of the people through existing private associations as well as through *ad hoc* voluntary associations established in connection with the administration of the agency's program.

3. *The decentralized administrative agency is given a key role in coördinating the work of state, local, and federal programs in its area of* {29} *operation; and a regional development agency should be given primary responsibility to deal with the resources of the area as a unified whole.* The place for the coördination of programs is in the field, away from the top offices which are preoccupied with jurisdictional disputes and organizational self-preservation. Coördination should be oriented to the job to be done, centering federal authority and its administrative skills and power upon the special needs and problems of the area.

These minimum essentials may be more clearly understood if they are examined as (1) the concept of managerial autonomy, (2) the partnership of TVA and local government, and (3) the ideal of basing unity of administration upon the natural unity of a region as an area of operation in resource development.

MANAGERIAL AUTONOMY

The TVA itself, as an autonomous administrative agency, "a corporation clothed with the power of government but possessed of the flexibility and initiative of a private enterprise,"[10] embodies the first of the essentials of decentralized administration: the freedom to make significant decisions on its own account. This freedom from centralized control is in turn a condition for the realization of the second essential, for the decentralized agency must be able to make its own choices, on the basis of its experience in the field, in order to maximize the "participation of the people themselves."

To this end, the TVA has demanded managerial autonomy[11] in its relation to the federal government. The Authority has insisted that if effective decentralization is to be achieved the organization must be permitted to retain real administrative powers unhampered by the administrative controls ordinarily exercised over government departments. Considered broadly, this demand has a wider significance than the establishment of the conditions of decentralized administration. It is also intended as revolt against the conception of government as a necessary evil and government officials as inherently tainted. In the eyes of some TVA officials, the stringent controls over personnel and financial policy normally exercised by civil service and budgetary agencies represent a {30} cultural lag, still attuned to a governmental structure limited in scope and essentially parasitic rather than playing a significant positive role in social and economic life. The rationale of strong housekeeping controls seems to be weakened rather than strengthened by the conditions of the new state with its ever broadening functions. A new problem arises: not one of restricting the powers and functions of government, but of developing new techniques which will permit, within the framework of positive government, the exercise of initiative and the kind of independence which can forestall the rise of centralized bureaucracy. Can the benefits of anticorruptionist controls be obtained outside of the old administrative devices? This question, in the opinion of the Authority's leadership, can now be answered positively.

It is doubtful that the experience thus far accumulated is sufficient to permit any exhaustive listing of the administrative freedoms necessary for that degree of managerial autonomy required by administrative decentralization and yet consistent with a reasonable measure of broad control by the President and Congress. But it is possible to state three such conditions of autonomy which are considered fundamental by TVA.

1. *Freedom from control by the Civil Service Commission.*—The Authority has been free of control by federal civil service laws from its inception, and its continued exemption from the Ramspeck Act of 1940 for the general extension of civil service has underlined its unique position. The TVA Act of 1933 provided for the establishment of an internal

civil service system under the control of the Authority, by stipulating, in Section 6:

> In the appointment of officials and the selection of employees for such operation and in the promotion of any such employees or officials no political test or qualification shall be permitted or given consideration but all such appointments and promotions shall be given and made on the basis of merit and efficiency. Any member of said Board who is found by the President of the United States to be guilty of violation of this section shall be removed from office by the President of the United States. And any appointee of said Board who is found by the Board to be guilty of a violation of this section shall be removed from office by said Board.

It is a well-known feature of TVA's administration that this provision was taken to heart and translated into an effective policy from the beginning with the vigorous support of Arthur E. Morgan, the first chairman of the Board. In recruitment, classification, and labor relations the Authority has been widely praised as a model of progressive personnel policy. There is some doubt whether control by the Civil Service Commission in the early days of the Authority would have permitted {31} the kind of experimental approach which was instituted. Since the merit system in the Authority is secured by its basic Act, and ten years of experience have indicated the feasibility of such a system under autonomous control. TVA maintains that the solution it offers to the problem of centralization upholds high standards of civil service as well as the principle of managerial autonomy. In the Authority's view the maintenance of such standards does not require the sharing of administrative control over personnel policy with another agency, particularly one located in Washington.

The special character of TVA's requirement of freedom from civil service controls—although not, to be sure, from the principle of merit and efficiency—was evident in the Authority's early history and was derived not only from its attempt to develop an experimental personnel program but also from its necessity to be free to choose a personnel adaptable to its policies. This required flexibility in recruitment, unhampered by regulations which assume that all who reach a certain grade on an examination are equally fit for a position. Of considerable importance in this connection was that TVA had to recruit in a situation of struggle. The Authority was subject to bitter political and legal attack from its inception, and required local control over personnel recruitment in order to avoid the employment of hostile men for key administrative posts. This was one of the points made by Senator Norris in opposing the extension of civil service to TVA under the Ramspeck Bill:

> Now, Mr. Chairman, here is another thing I have advocated, that I believe in. It is something which hasn't been presented even by the TVA although I have talked to members of the TVA Board about it. Without

referring to anyone in particular, I want to say that especially while a period of construction in TVA is going on, a period which will extend at least three years more, we never should make it possible for a spy of what I have denominated the power trust, through civil-service regulations, to get into the service of the TVA. That is another reason why I contend they have a present civil service far superior to that of the ordinary Government civil service.

It would be a comparatively easy matter for a group of men or women to pass civil-service regulations and examinations and get appointments in the TVA, where they would have access to the confidential files and confidential communications and discussions that take place. For instance, while those great lawsuits were pending, when millions in money would have been forthcoming in the twinkling of an eye to get information on the confidential actions and communications that were going on within the body of the TVA, there might have been great injury and sabotage.[12] {32}

The advantage of autonomous control over personnel policy, safe-guarded by the merit and efficiency section of the Act, lies in the ability of the agency to choose the kind of men it wants, on the basis of an implicit theory of a close relation between personnel and policy, and thus to shape the character of the organization in a unified direction. The organization may then rely on the weight of personality and group attitudes within its structure for the implementation of policy rather than solely on administrative rules. Of course, all organizations will tend to educate the administrative ranks after recruitment, with vary-ing success, in order to achieve a unified approach to questions of policy and consequently a loyalty to the decisions of the leadership. But civil service autonomy provides the opportunity of choosing personnel at the outset on the basis of an estimate of the kind of men needed for the effectuation of a given policy. It has the added advantage, not less important, of permitting the flexible adjustment of personnel policies to the kinds of administrative arrangements deemed especially ap-plicable to the local situation within which the agency must operate.

The relevance of the demand for managerial autonomy in civil service matters to the Authority's policy of coöperation with local in-stitutions was recognized by the Joint Committee which investigated the TVA in 1938:

A further consideration of great weight is the relationship between the Authority and various other organizations, public and private, elsewhere discussed in this report. The Authority has been able to coöperate with other governmental agencies, Federal, State, and local, sometimes by providing personnel or the means to be used in hiring personnel. No evidence of corruption or inefficient use of this power has been submitted. It has served as a device for obtaining desirable results without arbitrary use of Federal power and with a maximum of democratic participation. The Civil Service regulations would have to be carefully limited to avoid serious interference with this form of coöperation.[13]

The Authority itself has emphasized the importance of an autonomous civil service for a regionally decentralized administration. In opposing the application of the Ramspeck Bill to its organization, the TVA protested that the character of the Authority demanded that "managerial decisions should be made in the Valley, where the work was to be done, not in Washington."[14] {33}

> A large part of the work of the Authority is, moreover, carried on in coöperation with State and local agencies and with other Federal agencies.
>
> The Agricultural Extension Service, the county-agent system, the staffs and facilities of State universities and departments, the municipal and county officials and agencies throughout the Tennessee Valley, are all essential elements in the Authority's program. It is through these agencies that the Authority sells its power, distributes its fertilizer, encourages improved land practices, promotes research in industry and agriculture, contributes to health and related developmental programs. The Authority is the only regional agency of its type in the country, and its relations with other public agencies are therefore different in quality and importance from those of any other Federal agency. Its staff must be specially qualified to carry on a regional program through regional collaboration and participation. They are selected not only for general administrative experience in organizing such coöperative programs, but also for their special familiarity with regional problems, regional procedures, and regional agencies.
>
> There can be no genuine regional administration of Federal functions in the Tennessee Valley if the responsibility for so important a phase of the Authority's program as the selection of the men who carry on its work, and the determination of the terms and conditions of their employment, is lifted from the discretion of the board under the terms of its comprehensive statute and lodged in a separate agency with headquarters in Washington. Such action would mean the beginning of piecemeal disintegration of a regional administration. . . .[15]

The final sentence of the above quotation is actually the key. For although it might be shown that many of the concrete special problems faced by TVA as a regional agency could be solved within the framework of the federal civil service system, abdication of control over recruitment would still be inacceptable to the Authority. It is the denial of the principle of managerial autonomy implicit in the federal civil service which is decisive, and the threat that (a) federal personnel control would be only the first of a series of inroads on the unique privileges of the TVA within the federal system; or even if this were not true, (b) a change in the policies of an outside Washington agency which TVA cannot control might thus overturn existing arrangements in the Valley or hinder further regional adaptation.

2. *Freedom from control by the General Accounting Office.*—Under the provisions of the Budget and Accounting Act of 1921, the Comptroller General of the United States—an agent of the Congress in its exercise of the power of the purse—was given the responsibility of assessing the legality of expenditures by federal departments. This power is embodied in a procedure which makes it possible for the Comptroller General to disallow expenditures by refusing payment when a particular transaction is deemed illegal. {34}

From the viewpoint of managerial autonomy, control by the General Accounting Office brings to bear upon the government agency the general housekeeping laws used as criteria for the legality of the transactions under scrutiny. Among examples of such statutes are "those requiring that purchases of materials, with certain exceptions, be preceded by advertisement for competitive bids; that needed articles be purchased from the federal prison industries if available from that source; that purchase of land on which is to be erected any public building be preceded by a favorable ruling of the Attorney General as to the validity of the title; that printing and binding be done in the Government Printing Office; that law books and periodicals not be purchased with appropriated funds unless their purchase has been specifically authorized by statute; that no payments be made to any publicity expert except from funds specifically appropriated for that purpose."[16]

The Tennessee Valley Authority has claimed freedom from these statutes, as well as from the authority of the Comptroller General to settle and adjust its accounts. This freedom was challenged early in the history of the Authority and resulted, after a considerable debate, in a clarification of the TVA Act by Congressional amendment. This amendment, enacted in November, 1941, placed the Authority definitely within the jurisdiction of the Budget and Accounting Act of 1921, but at the same time relieved it of any onerous consequences of that jurisdiction. The Authority was authorized to make final settlements of claims and litigation, and the Comptroller General was restrained from withholding funds "because of any expenditure which the Board shall determine to have been necessary to carry out the provisions of said Act."[17]

The TVA considers it fundamental to its status as a corporation engaged in public business and therefore subject, at least in large part, to judgment in financial terms, that it be permitted to make its decisions on a business basis. This is especially important in such matters as procurement and the settlement of litigation on an expedient basis. The criterion of business efficiency does not of itself necessitate safeguarding competition through advertisement for bids in procurement; but it is concerned with obtaining materials of suitable quality at the lowest cost and with the least delay. Thus the normal requirements of a business enterprise appear to argue, in the TVA view, for granting managerial autonomy to a decentralized regional authority with business functions. {35}

3. *Freedom to apply revenues to current operational expenses.*— Among the rights which the TVA has defined as crucial to its administrative existence has been that of freely disposing of its revenues from power and other sources in the interests of efficient management. Moneys received by the Authority are paid into a special fund in the Treasury, and may be applied to operating expenses as determined by management within the Authority. This, in addition to insuring continuity of operations of its electric power plant, has permitted the Authority to make budgetary determinations, such as those for travel expenses, substantially on its own account. The normal alternative in the federal departments is that itemized budget allowances are made by Congress; since moneys are available only from appropriations and for specified items, these allowances are not only relatively inflexible but precarious. It is the position of the TVA that its freedom from Congressional allocation and appropriation permits management to make adjustments in terms of needs as they arise and change, with an eye to over-all efficiency rather than to the terms of the budget.

This entire procedure was called into question by Senator Kenneth McKellar of Tennessee when he introduced a bill amending the TVA Act so as to require its receipts to be paid into the Treasury on the same basis as regular government departments. Although this amendment was not enacted, it was conceived as a basic threat to the integrity of the TVA organization, and provided the leadership with another opportunity to emphasize its demand for managerial autonomy. Lilienthal put the matter strongly:

> If the McKellar amendment were law, the TVA region would at a single blow be pushed back industrially 25 years. Since continuity of power supply would depend upon whether successive Congresses would appropriate funds from year to year, existing industries in the Tennessee Valley would be obliged to move out of the region to a place where power supply could be depended upon without question.... Long-term contracts have been made, in reliance upon the TVA's power to use its revenues from those contracts to insure continuity of service. It would undermine those solemn obligations to change the provisions of law which largely induced entry into those contracts. I am not over-stating the case when I say that the McKellar amendment would make scraps of paper of more than 140 contracts made in good faith with the U. S. A.[18]

While the question of the continuity of power operations has been stressed, it seems fair to say that equally important in the minds of the TVA Directors has been the problem of maintaining the Authority free {36} from onerous political control. They have wished to avoid the introduction of ulterior considerations of political expediency into the determination of matters which should be decided on a business basis. Thus, in a speech to an Alabama community at the time of the McKellar controversy, Lilienthal said:

We meet here as typical representatives of several million of Americans engaged in a unique undertaking. Our course in the future as in the past will be beset with difficulties and opposition. Let us here resolve that we will fight at the drop of a hat against any intrusion of predatory politics into this partnership. Let us continue to administer this joint undertaking on principles of sound management. The people of the communities which you represent join with communities in the other Valley states to cry "hands off" to anyone who seeks to substitute politics in our partnership for business management.[19]

Viewed in this way it becomes clear that the issue cannot be resolved by providing the Authority with a special fund to defray emergency expenses while leaving all regular expenses, such as those for salaries, travel, and information, subject to determination through the annual appropriation procedure. For in general it is believed that political expediency may at any time become the deciding factor which will deprive the agency of its needed funds quite apart from considerations of business efficiency. With such a danger always present it would be inevitable that the decisions of the agency itself would be made with a view to minimizing possible threats to its funds, with the consequence that management techniques would become tied to the problem of influencing a Congressional committee. With political threats to its finances removed, the TVA is free to make decisions on particular matters in terms of internal efficiency judged by over-all results.

The three points stated above—an independent personnel system, freedom from housekeeping controls of the General Accounting Office, and discretionary power to apply revenues to operating expenses—have been cited by TVA officials as fundamental principles of managerial autonomy. Such autonomy represents one of the main features of the grass-roots doctrine, having to do largely with the relations of a regional agency to the federal administrative system.

Managerial autonomy, it will be noted, is not confined to the requirements of the Authority in its role as a business corporation. Thus, as described above, the freedom from control by the Civil Service Commission {37} applies to the developmental aspects of the TVA, such as conservation and planning, as well as to its power business. The idea of business efficiency as a criterion of professional management is thus extended to traditional governmental functions.

THE PARTNERSHIP OF TVA AND THE PEOPLE'S INSTITUTIONS

After some difficulties and initial disagreements, but still very early in its history, the Authority defined its approach to coöperation with the agencies and institutions already existing in the Valley. The alternatives seemed to be two: either to take a line which assumed that the TVA itself could and should carry out its programs by direct action; or to accept as legitimate and efficient a method which would seek out and even establish local institutions to mediate between the TVA and the people of the area. It was felt that an imposed federal program would be alien and unwanted, and ultimately accomplish little, unless it brought together at the grass roots all the agencies concerned with and essential to the development of a region's resources: the local communities, voluntary private organizations, state agencies, and coöperating federal agencies. The vision of such a working partnership seemed to define "grass-roots democracy at work."

In the Authority's view, the fundamental rationale of the partnership approach is found in its implications for democracy. If the TVA can be "shaped by intimate association with long-established institutions,"[20] that will mean that its vitality is drawn from below. By working through state and local agencies, the Authority will provide the people of the Valley with more effective means by which to direct their own destinies. The TVA may then become more integrally a part of the region, committed to its interests and cognizant of its needs, and thus removed in thought and action from the remote impersonal bureaucracy of centralized government.

The moral dimension of the grass-roots approach has been emphasized many times. The methods of TVA are proffered as more than technical means for the achievement of administrative objectives. They include and underline the responsibility of leadership in a democracy to offer the people alternatives for free choice rather than ready-made prescriptions elaborated in the fastnesses of planning agencies. By 1936, Lilienthal had formulated the bases of such a policy. {38}

> This matter of making a choice available, which is the duty of leadership, seems to me critically important. There are two ways of going about many of these matters. There, for example, is a steep slope which has been denuded of trees by the farmer. He has to make a living. He needs this steep slope to grow the things that will keep his family alive, and so he cuts the trees down and plants his corn, and the soil is washed off in a few years, and the nation has been robbed of just that much of its capital assets.... Now one way of going about it is to say, "We will pass a law that any farmer who cuts down the trees and cultivates a slope steeper than a certain grade is incapable of farming. He is injuring the community and the nation, and by this law we will take his land away from him and turn it back into forest or meadow." That is one way.... Then there is the other method, which the TVA has pursued, of giving the farmer a chance to make a choice; recognizing that the farmer does not cut down those trees because he enjoys cutting down trees or because he likes to see the soil washed off and destroyed but because he has a problem of feeding his family and making a living. Give him a choice—a free choice—by making it possible for him to use his land in such a way that he will not only be enabled to support his family but at the same time protect that soil against depredation. This is only one illustration of many of this conviction I have, that a man must

be given a free choice, rather than compelling a choice or having super-men make the choice for him.[21]

In this way the Authority has applied a moral sanction to its program of giving the people and the existing institutions in the area a chance and the means to participate in an over-all program.

The orientation toward local agencies is also a product of the conception that the resources of a region include its institutions, in particular its governmental agencies. The Authority deems it part of its obligation in connection with resource development that these local governmental institutions be strengthened rather than weakened, that they be supplemented rather than supplanted. In doing so, the Authority directs its effort toward developing a sense of responsibility on the part of the local organs and, what is equally important, toward providing them with a knowledge of the tools available to put that responsibility into action.

The TVA's policy of strengthening local institutions is linked to its broad responsibilities for regional development. Thus Gordon R. Clapp, formerly general manager and now chairman of TVA, has emphasized that maximum regional development is a function not only of the physical resources of a region but of administrative or managerial resources as well.[22] This must be evident to all who understand the difficulty of {39} bridging the gap between a recognition of the needs of a region and the establishment of suitable methods for their fulfillment. Moreover, the plans of engineers and scientists depend for their effectuation upon the decisions and efforts of the people who actually operate upon the soil and other resources of the area. To attain the coöperation of these people is a matter not of simple exhortation, but of persuasion and organization, of practical attempts through the solutions of their individual problems to link them with the public goals of the TVA.

A list of agencies with which the TVA has maintained some form of coöperative relationship includes nearly all of the governmental institutions in the area: municipal power boards, rural electric coöperatives, school and library boards; state departments of health, conservation, and parks; state and local planning commissions, agricultural and engineering experiment stations, state extension services, and others. In developing these relationships TVA has applied the rule that "wherever possible, the Authority shall work toward achieving its objectives by utilizing or stimulating the developing of state and local organizations, agencies and institutions, rather than conducting direct action programs."[23] In addition, a number of federal agencies, notably technical bureaus of the U. S. Departments of Agriculture and the Interior, the Army Engineers, and the Coast Guard, have coöperative arrangements with the TVA. Notable also are the *ad hoc* organizations and conferences which have been established as vehicles for coöperation among the administrative agencies within the Valley. These include, among others, a semiannual conference of directors of extension services and of agricultural experiment stations of the seven Valley states, the U. S. Department of Agriculture, and the TVA; the Tennessee Valley Trades and Labor Council, bringing together fifteen international unions of the American Federation of Labor Building and Metal Trades; an annual conference of contractors and distributors of TVA power; and the Tennessee Valley Library Council. Such gatherings help to lay a sound foundation for regional unity, focusing the efforts of many agencies on the region as a central problem.

The form of coöperation with state and local agencies varies, but the pattern of intergovernmental contract has been most fully developed. Such contracts often include reimbursement by the Authority for personnel and other facilities used by the state in carrying on the coöperative {40} program. In many cases, the ideal outcome is viewed as the tapering off of TVA contributions until, as TVA's responsibilities recede in importance, the local agency carries on by itself. Thus the states have in some cases begun planning work through their own commissions with the material help of TVA; later, state funds have been secured with a view to continuing the work when TVA's responsibilities for the readjustment of reservoir-affected urban communities would terminate. In coöperating with the local governments, TVA attempts to establish a pattern which may be continued after TVA aid has ceased.

The objective of stimulating local responsibility among governments and associations within the area is basic to the grass-roots approach. But there are other reasons which support it as sound administrative policy. The existing facilities of the states, even though they may be inadequate, are used to capacity, thus avoiding the establishment of duplicate services and personnel with parallel functions. The TVA is not anxious to have its own men in the field and is willing to forego the prestige that comes from identification as "TVA men" of agents performing services paid for out of TVA funds. The staff is educated to feel most satisfied when it can show evidence that a local organization has carried on TVA work and been permanently strengthened by the experience and in the eyes of its public. In addition, utilization of existing agencies permits TVA to shape its program in conformity with the intimate knowledge of local conditions which such agencies are likely to have; at the same time it is possible to restrict the size of the Authority's direct working force.

The attempt to create a working partnership between the TVA and the people in carrying out a common program for regional development goes beyond the strengthening of existing governmental agencies, though this objective is vital. The meaning of the partnership is contained as well in the use of the voluntary association as a means of inviting the participation of the people most immediately concerned in the administration of the program. In this way, the farmer or the businessman finds a means of participating in the activities of government supplemental to his role on election day. If there is fertilizer to be distributed, farmers are invited, on a county and community basis, to participate in locally controlled organizations which will make decisions as to the most effective means of using that fertilizer in the local area. If government land is to be rented, a local land-use association is organized so that the conditions of rental can be determined with maximum benefit for the community. If power is to be sold in a rural area, a coöperative provides a {41} consumer ownership which retains profits in the community and makes possible a management guided by community problems and local needs. If the business area of a city must be modified because of newly flooded lands, let a locally organized planning commission work out the best possible adjustment of special interests and long-range planning goals. Thus, at the end-point of operation, the specific consequences of a federal program may be shaped and directed by local citizens so that its impact at the grass roots will be determined in local terms. This procedure is not only democratic and just, but undoubtedly adds measurably to the effectiveness of the programs, which will be conjoined to the special desires of those affected and thus have the benefit of their support and aid.

The policy of consciously working with and through local institutions is, in the Authority's view, integrally related to its relatively autonomous position within the federal system. It is precisely the flexibility accorded to the TVA management which has enabled it to keep in mind its broad concept of regional development and at the same time to seize upon whatever opportunities might arise to implement the concept concretely. Nationally directed restrictions as to employment of personnel, a host of regulations framed in national terms, would doubtless greatly restrict the ability of TVA to establish procedures attuned both to its substantive objectives and to the grass-roots methods by which they are carried out. It would surely inhibit the freedom to search out techniques uniquely adapted to the special situations of some particular state government or community if TVA did not have the power to make its own decisions and to take the initiative in fostering coöperative relationships. Moreover, the absence of discretion might well be psychologically decisive in hobbling the TVA staff by binding it to the customs and traditional modes of action laid down by the broader hierarchy into which it might be absorbed.

DECENTRALIZATION AND REGIONAL UNITY

While the problem of method is the nub of the TVA approach, the latter may not be divorced from the program which called it into being. The outlook of the TVA leadership stresses decentralization as vital not only for the future of democratic government but also as an indispensable tool for the effective development of regional resources. The TVA idea is also a regional idea which looks toward the formulation and execution of a unified program for resource development; and it is believed that responsibility and authority for the fulfillment of such a program should {42} be allocated to a single agency. This does not mean that a single agency will actually carry out all the necessary activities, but it does mean that it will provide over-all direction in terms of regional goals. It means that governmental initiative, localized and focused in an agency which has responsibility for unified development, will be available to seize opportunities as they may arise, and to make the most of them.

Any general extension of the TVA idea must plainly be rooted in a recognition of the diversity of needs and potentialities which characterizes the various sections of the United States. This diversity demands flexibility on the part of federal authority and the use of a method which can adjust a national program to the special problems of the area of operation. It is difficult and perhaps impossible to specify in any exact way how the lines may be drawn around an area to say this is a Region.[24] That will have to be done in terms of the existence of a problem area, or a resource base, or some other criterion or combination of criteria which will serve to reflect the practical unity which ordinary men understand when they associate together for regional objectives. What is important is not precise specification or neat boundaries, but the recognition of focuses of potentiality and need which, treated as wholes, can contribute most effectively to regional development and thereby to national welfare. Diversity in nature and tradition requires diversity in program and policy; for the idea of regional administration is frankly opposed to those forces which strive to introduce uniformity into all aspects of government programs.

The decentralized regional agency offers a means of creating a center of regional responsibility, planning, and coördination within the framework of the existing federal system. Through it, functions too broad to be undertaken by a single state and yet not actually national in scope may be initiated at the proper level and given direction and scope in terms of the special problems of an area, at the same time that national policy may be brought to bear upon regional needs. Such staff agencies as the Bureau of the Budget and a National Resources Planning Board can, as integral parts of the Presidency, direct from the national government matters of broad perspective. The decentralized regional agency is, however, not the same as the regionalization of national agencies through the establishment of field offices. Such federal outposts are not charged with responsibility for integral regional development; and, equally important, regional unity demands the continuous development of a program {43} based on study and decision in the field. A truly regional agency is multifunctional and single purposed. It has the means and the authority to engage in or to initiate several major and many lesser projects in accordance with the developmental needs of the area. At the same time, the outlook of the agency is integral, devoted to the unified and conservative exploitation of the region as a whole.

In the TVA, the idea of unified resource development is associated with a characteristic spiritual lesson. Largely under the influence of Dr. H. A. Morgan, former President of the University of Tennessee and one of the original members of the TVA Board, the agency's outlook has been framed in terms of the conception of a "Common Mooring," linking industrial and governmental activities to a fundamental human responsibility for the prudent utilization of natural resources.[25] "Dr. H. A.'s" lectures are well known in the Authority, and they are presumably a useful method of communicating the unified outlook to the staff. This Common Mooring consists of the basic factors provided by nature which make organized life possible: the inexhaustible elements in the air, water, and soil and the exhaustible resources which a morally guided intelligence must husband:

> It has not been an easy task. . . for nature throughout the millions of centuries to keep the elements and compounds within the evolutionary program so that when man should appear he would find a universe of complex interacting interdependent elements and compounds at his service. These have been so governed by laws that man's environment has permitted not only his existence, but that of all plants and other animals. This complex has definitely challenged man to discover his relation to his Creator and the creative process through his relation to his environment and to the mineral, plant, and animal kingdoms in which he found himself. This challenge has not stopped here. Over the centuries there has developed within man a sense of responsibility and moral obligation to assist nature in augmenting, through intellectual concepts not available to any other beings, the program that brought him into being.[26]

The TVA program is viewed as an application of the Common Mooring to regional development. It is a "single-purpose approach to many-sided problems," treating as fundamentally interrelated the basic, inexhaustible resources: air, water, and land. The unified approach made possible by a single agency is a needed antidote for the division of the universe made necessary by specialization. Put another way, TVA makes for {44} catholicity of outlook, for the leadership of the single agency acts as an integrating force in educating the component parts of its staff to view the problem of resource development as a whole.

The existence and the program of a regional agency is tied to the definition of a resource base. This will not be the same for all areas, but presumably it is possible to specify some key resource whose development can be the hub of a general program for each area. The decentralized agency should be made directly responsible for the development of the key resource, and be given the authority to make such arrangements with other governmental agencies as will most effectively fulfill the over-all requirements of the region. With these powerful tools, the regional agency can become an effective coördinator, mobilizing all possible participants in the interests of an integral plan.

A regional agency is the reflection of the physical unity of the resource base of the region. In the Tennessee Valley, the control of water in the river channel is bound to the control of water on the land. The river requires a unified system of dams and reservoirs; the land requires the development of sound agricultural practices and the wider use of phosphatic fertilizers. These jobs require different techniques and specialized personnel, but the development of the area demands that they be seen as a whole, as they affect each other. A regional agency can undertake to bring those perspectives together; it can also take into account the social consequences of its programs as the TVA does when it aids communities to intelligently readjust themselves to newly created reservoirs, and flooded-out farmers to relocate. The interrelation of social phenomena makes for a spreading network of consequences following in the wake of action; as a result, a special public emerges. It is this public, plus the national interest ultimately involved, which forms the political basis for the decentralized structure of the regional agency.

So runs the official doctrine.

◊ ◊ ◊

It will be noted from the above account that the concept of administrative decentralization or grass-roots administration, as used by TVA, refers to three administrative levels:

1. The location of administrative control in the area of operation, with the Authority as a whole, in relation to the federal government, taken as an example.

2. The carrying on of operations with and through existing institutions already organized in the area of operation. The relation of the {45} Authority to the agricultural extension services of the land-grant colleges is one of the important examples of this procedure.

3. The participation of local people at the end point of administration of the program, for example, through county soil improvement associations set up in connection with the TVA fertilizer test-demonstration program.

In the following pages an attempt is made to subject the official doctrine to functional analysis, and to set forth in detail the operation of one important phase of the grass-roots procedure, tracing its consequences for the role and character of the TVA. The major direct concern of the study is with level 2 noted above, secondarily with level 3. However, it will be evident in the course of the analysis to follow that the policy pursued by TVA in its own area cannot be free of consequences for its relation to the federal government, and therefore for level 1.

[1] From a mimeographed transcript of remarks by David E. Lilienthal to the Conference on Science, Philosophy, and Religion, Columbia University, New York (August 28, 1942).

[2] David E. Lilienthal, "The TVA: A Step Toward Decentralization," address before the Graduate Faculty of Political and Social Sciences, New School for Social Research, New York City, April 3, 1940 (mim.).

[3] David E. Lilienthal, "The TVA: An Experiment in the 'Grass Roots' Administration of Federal Functions," address before the Southern Political Science Association, Knoxville, Tennessee, November 10, 1939 (Knoxville: TVA Information Office, no date), p. 4.

[4] Ibid., p. 7.

[5] Alexis de Tocqueville, Democracy in America (New York: Knopf, 1945), I: 86.

[6] Ibid., p. 87. Whether Lilienthal has been correct in resting his argument on De Tocqueville's analysis is not here in question.

[7] Lilienthal, "The TVA: An Experiment...," p. 10.

[8] David E. Lilienthal, "Management—Responsible or Dominant?" Public Administration Review, I (Summer, 1941), 391.

[9] Loc cit.

[10] Message of the President, 10 April 1933. House Doc. 15, 73d Cong., 1st sess.

[11] This study is concerned primarily with the TVA's relation to local institutions, as expressed in the implementation of the grass-roots policy. The problem of managerial autonomy as such—concerning the relation of the Authority to the federal government—is not directly relevant. However, space is allowed here to the official position on managerial autonomy because of TVA's insistence that precisely such autonomy is indispensable to a proper execution of a grass-roots policy. It has therefore seemed desirable to discuss autonomy in detail, since this chapter as a whole is devoted to an exposition of the TVA viewpoint.

[12] Hearings before the Committee on Civil Service, U. S. Senate, on H. R. 960, "An Act Extending the Classified Civil Service," 76th Cong., 3d sess. (1940), pp. 193-194.

[13] Report of the Joint Committee to Investigate the Tennessee Valley Authority, Senate Doc. 56, 76th Cong., 1st sess. (1939), p. 60.

[14] "Statement on Application of H. R. 960 to the Tennessee "Valley Authority," Hearings before the Committee on Civil Service, U. S. Senate, 76th Cong., 3d sess. (1940), p. 249.

[15] Loc. cit.

[16] David E. Lilienthal and Robert H. Marquis, "The Conduct of Business Enterprises by the Federal Government," *Harvard Law Review*, 54 (February, 1941), 578.

[17] 55 Stat. 775-776 (1941). The Authority's corporate freedom in this respect was preserved under the provisions of the Government Corporation Control Act of 1945. See 59 Stat. 597 (1945).

[18] *Hearings* on S. 2361, Committee on Agriculture and Forestry, U. S. Senate, 77th Cong., 2d sess. (March, 1942), pp. 175-176.

[19] David E. Lilienthal, "The Partnership of Federal Government and Local Communities in the Tennessee Valley," address before gathering of community leaders of northern Alabama, sponsored by Decatur Chamber of Commerce, Decatur, Ala., July 30, 1942 (mim.).

[20] H. A. Morgan, Chairman, TVA, "Some Objectives and End Results of TVA," address before annual meeting of the American Society of Agricultural Engineers, Knoxville, Tenn., June 23, 1941 (mim.).

[21] David E. Lilienthal, transcript of TVA lecture, Knoxville, Tenn., June 12, 1936 (mim.).

[22] Gordon E. Clapp, "The Administrative Resources of a Region: the Example of the Tennessee Valley," in *New Horizons in Public Administration* (University, Ala.: University of Alabama Press, 1945), chap. IV. The publication of this volume, first of the University of Alabama Press, is itself an indication of the TVA's attempt to strengthen local institutions. In conjunction with TVA, the universities of Alabama, Georgia, and Tennessee launched a Southern Regional Training Program in Public Administration which provided internships for ten college graduates in 1944. The series of lectures published in *New Horizons* was a feature of the training program.

[23] As formulated at an Administrative Conference of TVA staff members, April 21, 1943.

[24] Cf. D. E. Lilienthal, *TVA: Democracy on the March* (New York: Pocket Books, 1945), p. 167.

[25] See Ellis F. Hartford, *Our Common Mooring*, prepared for the Advisory Panel on Regional Materials of Instruction for the Tennessee Valley (Athens, Ga.: University of Georgia Press, 1941).

[26] H. A. Morgan, "The Common Mooring," in *Applications of the Common Mooring*, compiled and edited by Howard P. Emerson, Chairman, subcommittee on Presentation of the Common Mooring, Advisory Panel on Regional Materials of Instruction for the Tennessee Valley, Knoxville, Tenn., June, 1942 (hectographed), p. ii.

CHAPTER II

THE FUNCTIONS AND DILEMMAS OF OFFICIAL DOCTRINE

The grass-roots policy is merely a rationalization. It is absurd to have a federal agency trying to work through organizations which place it at a second remove from the people.

A TVA STAFF MEMBER (1943): AN EXTREME VIEW

THE PARAPHRASE of official doctrine, briefly stated in the preceding chapter is not, of course, complete. Ideas sincerely held and highly prized are not likely to be acceptably presented by a mind alien or even detached. The leadership of the TVA would be quick to point to subtleties unnoticed, ramifications ignored, and meanings crudely put. Such an attitude, if expressed, would be readily understandable. For this is a doctrine to which the TVA leadership is deeply committed. The idea of a grass-roots administration is, as we have already noted, no casual or minor element in the consciousness of the Authority's staff. It is, on the contrary, one of the symbols most frequently referred to inside the organization. Speeches to new employees, letters to information seekers, a popular book written by the chairman, memoranda discussing projects and procedures, all attest to the importance of the grass-roots idea in TVA. The systematic promulgation of these ideas helps to define the character of TVA as an organization and serves to shape the outlook of its staff. Open though it is to various interpretations, there can hardly be any question that, at least in the regional development departments, the grass-roots approach is accepted without question as official policy and for the most part is warmly endorsed as effective administrative procedure.

SOURCES AND FUNCTIONS

Among the many and pressing responsibilities of leadership, there arises the need to develop a *Weltanschauung,* a general view of the organization's position and role among its contemporaries. For organizations are not unlike personalities: the search for stability and meaning, for security, is unremitting. It is a search which seems to find a natural conclusion in the achievement of a set of morally sustaining ideas, ideas which lend support to decisions which must rest on compromise and restraint. Organizations, like men, are at crucial times involved in an attempt to close the gap between what they wish to do and what they can do. It is natural that, in due course, the struggle should be resolved in favor {48} of a reconciliation between the desire and the ability. This new equilibrium may find its formulation and its sustenance in ideas which reflect a softened view of the world. The ethic of revolt, of thoroughgoing change, assumes that human and institutional materials are readily malleable and that disaffection from current modes of thought and patterns of behavior can be long sustained in action. But leadership must heed the danger of strain and disaster as recalcitrance and inertia exert their unceasing pressures; in doing so, it may see wisdom in the avowal of loyalty to prevailing codes and established structures. The choice, indeed, may often lie between adjustment and organizational suicide.[1]

This quest for an ideology, for doctrinal nourishment, while general, is uneven. Organizations established in a stable context, with assured futures, do not feel the same urgencies as those born of turmoil and set down to fend for themselves in undefined ways among institutions fearful and resistant. As in individuals, insecurity summons ideological reinforcements. The TVA was particularly susceptible to this kind of insecurity because it was not the spontaneous product of the institutions in its area of operation. Thus one Southern critic, Donald Davidson, wrote:

> The regional psychology of the TVA conception is not native to the South. The TVA was not created in response to a Southern crusade—although some Southern Congressmen had long agitated for a final disposition of Muscle Shoals. There was no popular outcry in the South for a TVA. The sponsor of the TVA Act was a Middle Westerner, Senator Norris, who was crusading against the electric utilities.... The five states concerned had no opportunity to debate the project or to contribute ideas or leading personnel, or by any direct means to make known their opinion, if they had any; the project was superimposed (however benevolently) upon them.[2]

The often-remarked change in attitude toward TVA among people of the Valley—from hostility or suspicion to acceptance and support—lends weight to Mr. Davidson's observation. In 1946 a reporter for the Denver *Post* wrote from Knoxville:

> The TVA is such a revolutionary type of government organization that its entrance into this generally conservative, middle-south region thirteen years ago was marked by many bitter controversies, the wounds of which, however, have mostly healed now. {49} From small business-men, school teachers, farmers, newspaper writers, city officials, railroad workers, and professional men, I have gained the impression that those who still oppose TVA here are in the minority, decidedly.... A Knoxville businessman: "This is a conservative community. This is the heart of the Bible Belt. Our people voted down Sunday movies, for example. When TVA set up headquarters here with flocks of 'bright young men' many of whom were not yet dry behind the ears, our people resented it and

referred to them as 'those foreigners.' Now the resentment is gone. The socially minded theorists in TVA who couldn't deliver in actual practice have gone."[3]

The TVA could not take its position for granted. It had to feel its way. This required the formulation of a policy that would reassure external elements and would so educate its own ranks as to maximize the possibilities of social acceptance.

But adjustment is a complex process. In pursuing an analysis of this sort it is essential to remember how diverse and unstable are the sources and functions of any set of guiding ideas with which an organization becomes identified. Some organizations are born of ideological conflict, and administered by the victors with a view to furthering "the cause." If so, the doctrine espoused by the leadership may be understood only with reference to the symbolic status of the organization. Such might have been true in TVA if planning as such or positive government or some other component of the doctrine of the liberal-socialist movement in the United States had been emphasized by the directors as the special TVA philosophy. Then the problem of theoretical interest might have been to examine the possibilities of a gap between what men said (in the light of doctrinal commitments) and what men did (under the pressure of practical exigency, the force of circumstance). But there are other sources of ideas, sources close to the special problems of organization as such, which require some attention.

One such source of official doctrine derives from the simple fact that formal organizations live in institutional environments. Each is a coöperative system, and involves its members wholly, with all the cultural predispositions and tangential commitments given them by lives apart from the formal purpose of the organization. Thus the sheer weight of special education provided by the community from which personnel is drawn will shape and limit the ideas which can be developed by the organization. Moreover, there are specific local pressures, especially of the various client groups served, which infuse the organization with new goals and interest-defensive ideas. {50}

In addition, ideas arise from the need for internal communication as a means of developing organizational unity and homogeneity. This functions both to present a consistent interpretation to the world, especially when subject to attack, and to establish a foundation for the smooth flow of directives, the ready acceptance of executive decision. This unity, when based on the cultivation of an appropriate set of attitudes, provides a framework for the development of special policies which will be attuned to the basic view of the leadership. The organization takes on the aspects of a social organism, with a set of precepts that will be taken for granted. This has useful administrative consequences:

1. If a basic point of view is laid down and integrated into the psychology of the second and third level leadership, if not farther down

the ranks, the possibilities of decentralization without injury to policy are vastly enhanced. It is almost axiomatic that in new organizations, in which the adherence of the administrative personnel to the viewpoint of the directorate is not dependable, measures of formal control from the top must be introduced. When, however, indoctrination of official policy has been sufficiently extended, formal controls may be loosened. Ideas and the attitudes they foster may serve as surrogates for a system of rules and formal discipline. Thus in the early days of the TVA, before the adoption of the Employee Relationship Policy in 1935, a Labor Relations Division handled grievances directly with employees. An employee could by-pass the supervisory hierarchy in his own division and contact "Labor Relations" which had the power to investigate and force settlement. This control through an organ of central management was considered necessary because in the beginning, according to a staff member, TVA had "many supervisors, high and low, fresh from the notoriously antilabor construction industry. Hence the wielding of a big stick was thought necessary."

2. The educational function common to all such structures of painlessly but effectively shaping the viewpoints of new members is facilitated, thus informally but effectively establishing the rubrics of thought and decision. This is well understood in practice, reflected in the use of organizational labels (a "Forest Service" man, an "agriculturist," etc.) so that special attitudes and characteristic administrative methods may be identified.

3. A consistently developed doctrine providing a sense of unity to the organizational endeavor is particularly useful in binding together technical experts who otherwise tend to emphasize their special professional interests. Formal methods of coördination may be buttressed by the communication of a unified outlook, for the latter will tend to make the {51} technicians think in terms of the organization as a whole. In the TVA, the problem of coördinating the views of experts has been considered especially important by central management. The ability of a representative of one division to appear before the Board of Directors representing not only his own but some other division is hailed as a significant achievement. Such a welding of viewpoints is aided by the coördinating power of a well-developed set of ideas.

4. A further source of organizational doctrine is the normal tendency for an agency to defend its own functions. To some extent, even the leader of a subordinate division is expected to speak for and defend the interests of his division. As a matter of fact, his role is defined by the system as one of representation, not without reason. It is convenient and perhaps necessary for a central coördinating agency to circularize its staff components for opinion as to policy; the significance of such opinions lies in setting forth the special needs and outlooks of the constituent divisions. Coördination requires such a collation as the raw material for decision that will take all factors into account. Moreover, the possibilities of competition among subordinate sections are thereby

furthered; this has an incentive value and to some extent is an indication of the vigor with which a program is being pushed, although degeneration into bureaucratic politics may and does occur. The sub-leader must also see that the needs of his staff are represented when decisions affecting it are made,[4] a function which will turn his attention inward to emphasize his organization as a value in itself. Professionally, too, the personnel becomes identified with the agency, deriving prestige and disesteem from the fortunes of the organization itself. To defend the organization is often to defend oneself. These defensive activities are aided when a set of beliefs is so fashioned as at once to fortify the special needs or interests of the organization and to provide an aura of disinterestedness under which formal discussion may be pursued.[5] {52}

The considerations just noted refer to the role of any official doctrine in an organization. However, in pressing our inquiry into the functions of the TVA's grass-roots doctrine, it is necessary to examine its origins, in particular whether the doctrine arose from within the Authority or was provided for it at the outset. This is important for an analysis which attempts to explain the function of the doctrine by the special needs of the organization as an adaptive structure.[6]

The elaboration of the grass-roots policy as official organizational doctrine seems to have filled what amounted to an ideological vacuum. The formulation of this doctrine did not precede the establishment of TVA, nor did its precepts materially influence the nature of the organization created by the TVA Act. The form of the TVA organization was related more to the specific political situation at the time of its inception than to any broad principles of public management. To be sure, the advantages of a corporate organization having the powers of government and the flexibility of private enterprise—a form already familiar in American government—were apparent; but the key elements of the situation in 1933 which seemed to point to the need for an autonomous agency were (1) the struggle over public power, and (2) the experimental character of the TVA as a planning agency. Neither implied a self-conscious doctrine such as that developed by the Authority.

1. An autonomous agency, having broad powers of discretion, was an appropriate device for organizing a battle against the electric utility interests. A leadership could be chosen which understood its responsibilities in terms of the national struggle undertaken by the Roosevelt administration. The autonomous directorate could muster its forces freely, according to the needs of the conflict, without being bound by rules formulated for extraneous ends and without being hampered by the customs and prior commitments of a federal department. Thus we have already noted (see p. 31 above) that the demand to be free of U. S. Civil Service Commission control was supported by Senator Norris because of his fear that "spies" of the "power trust" might infiltrate into {53} the organization and cause "great injury and sabotage." Again, flexibility allowed the TVA in the use of its funds was justified partly on the

basis of the electric power controversy. This flexibility would make it possible for the agency to seize opportunities as they arose for advancing the cause of public power. For example, a local electric coöperative whose area was threatened with invasion by private utilities might be given aid by the Authority without undue hindrance or delay.

2. The TVA Act, and more fully the message of President Roosevelt requesting the legislation, represented a political challenge, even if it was not an entirely new departure in the exercise of federal authority for general welfare objectives. According to the President, the TVA was to be "charged with the broadest duty of planning for the proper use, conservation, and development of the natural resources of the Tennessee River drainage basin and its adjoining territory for the general social and economic welfare of the Nation."[7] In the public mind, the meaning of TVA was bound up with the values of planning and social responsibility as corollaries of the public-power controversy. This represented a model of positive government which could not be put forward for the federal government as a whole but rather, for reasons of strategy, as an experiment to be extended as conditions might warrant. In that capacity, and separated from the existing federal apparatus, the agency would remain in the national spotlight as a living example of what planning could accomplish. Moreover, the Authority still remains the receptacle of the hopes of those who stand for national planning; its autonomy is defended in political terms, on the basis of its program rather than the grass-roots method.

It seems clear that in 1933 the implications of the managerial form of the TVA for the future of federal administration were not well understood. Certainly the problems of public power and planning were pressing, and no systematic doctrine, such as the grass-roots theory later put forward by the Authority, was envisaged. It is true that the Authority was authorized to coöperate with local and state institutions, but no positive program of strengthening existing institutions was outlined as part of the conception of the TVA. The Authority was also authorized to coöperate with federal agencies, but this has not been stressed by TVA as a basic objective; Lilienthal has suggested that regional agencies should be compelled to coöperate with local and state agencies but has not recommended a similar injunction with respect to organs of the federal departments. In general, TVA is willing to work *with* federal {54} agencies, but emphasizes working *through* state and local agencies. In relationships with federal agencies, the grass-roots policy is pressed: while there is little difficulty in getting along with federal agencies which work through established state and local channels, where they do not, TVA considers that it has a responsibility to "make a point of method, trying to persuade the federal agency to modify its method."[8] This special point of view was developed by the Authority itself, and in its systematic form represents a unique approach to the administration of federal functions. There is some indication, moreover, that the TVA theory was not accepted as an administrative pattern by President Roosevelt. In his message of June 3, 1937, in support of the Conservation Authorities Bill, the President specified that "projects authorized to be undertaken by the Congress could then be carried out in whole or in part by those departments of the Government best equipped for the purpose, or if desirable in any particular case by one of the regional bodies."[9] This clearly did not accept the grass-roots approach of the TVA, which calls for the decentralized execution of federal programs as a matter of fundamental policy.

If, as has been suggested, the special administrative form of TVA was derived from the circumstances of its inception rather than from general principles, it should be easy to see that an ideological vacuum would arise soon after the initial period. After victory was assured, and it became clear that the TVA was to endure, it was necessary for the leadership to develop some notion of its place within the American political system. Somehow an answer had to be found to the question: What now? Is the Authority a viable method of administering federal functions on a permanent basis, apart from the context of conflict and experiment? It was natural—though perhaps not inevitable—that the leadership should look with pride at the organization which had won so much praise, and natural that it should feel that the TVA method had an enduring value as effective management in a democracy. To some extent, fondness for the grass-roots theory resulted because it was raised as a banner in the internal conflict which rocked the agency not long after its inception.[10] A unified outlook was not known until the removal of Arthur E. {55} Morgan by President Roosevelt. The struggle which led to that removal is often interpreted as having involved the question of whether the TVA would work with the people of the Valley or impose its program despite them. However that may be, it is clear that originally there was no unity, and that the special outlook of Harcourt A. Morgan and David E. Lilienthal was formulated in the course of controversy. The latter is always an effective environment for the crystallization of loyalties to a certain set of ideas.

The basic functions of the official TVA doctrine have been (1) internal, satisfying such needs as effective communication, and (2) adjustment to the area of operation. The content of the doctrine is not, perhaps, of great significance for the former, but it is decisive for the latter. Whatever the specific formulation, the idea of taking account of, or working with, or otherwise accepting the existing social institutions of the Valley is clearly consonant with the iron necessities of continuing to exist and work in that area. Whether or not the TVA had developed the grass-roots theory, it would have found it necessary in some way to get along with those forces which might wreck the program if moved to resistance. In relation to the states, counties, and other local agencies, the TVA could not have operated successfully without framing its program within the existing pattern of government, including the powers and the traditional prerogatives of the local units. That must be taken for granted by any organization, public or private, which seeks to accomplish a large-scale program within a populated area. The alternative is force—an alternative which even occupying armies in enemy lands are loathe to use. In the field of economics, too, the TVA Commerce Department must assume that the main industrial base in the Valley will be private industry. Its work must be based on that fact, so that it may convince existing industry of the value of good stewardship of the Valley's resources. Thus in one sense the grass-roots approach simply verbalizes an administrative approach which any agency would follow of necessity.

The formulation of a doctrine placing a halo over procedures which might in any case be considered merely normal and necessary, functions to clear away the suspicions and latent resistance with which an agency imposed from above might be greeted. Such a doctrine also provides a vehicle for the justification of special efforts which may be made to placate the local leaderships and give them a stake in the fortunes of an organization not their own. Thus frontal appeals by local government to the TVA for favors may be avoided if the Authority itself systematically {56} exploits opportunities to assist these units with advice and money. The theory serves also as a means of organizing a "mass base" for the Authority by formulating and justifying aid to labor unions, coöperatives, farmers, farm organizations, universities, and state governments. Such a base serves as a protection if the agency is subject to attack. Though emphasizing again that we do not discount the sincerity of the TVA leadership, it is clear from the record that the Authority expects support from those agencies with which it works. Thus when, in 1941, the President of the University of Tennessee publicly criticized the proposed construction of Douglas Dam, and seemed to call into question the general benefit of TVA for the Valley, the Authority was much displeased, and threatened to "reconsider" the method by which it carried on its agricultural program through the land-grant colleges. When it is understood that the method in question is offered as one of the prime examples of the grass-roots approach, the significance of this can easily be appreciated.

The official doctrine does not appear to fill all of the TVA's needs for adjustment. Among the most important of the gaps recognized by members of the staff is the need to adjust the organization to the existing structure of the federal government. That, too, is part of the Authority's environment, though during the tenure of President Roosevelt the need to adjust the TVA to the federal structure may not have been as important for the survival of the organization as adjustment to local institutions. At least in terms of the development of TVA doctrine, the need of adjustment to the area of operation has been taken as basic. For the TVA has failed to develop a formulation which would be acceptable to the administrators in Washington, and in a practical sense there are areas of maladjustment which remain unreconciled. One TVA staff member has said that "when TVA was getting started it was perhaps necessary to avoid getting sucked into the federal stream, but now the Authority must develop effective working relationships with the old-line agencies.... TVA's relations to federal agencies have produced some enemies in the administrative setup, throwing anti-TVA barbs in the fight in Congress." Relationships with some long-established agencies in Washington are very good, but it is clear that the TVA's role in the federal structure has not been satisfactorily formulated. A Washington "bureaucrat" has written:

> . . .most of the discussions concerning possible forms of organization for resource development have largely ignored the possible effects of projected forms of organization on the general governmental framework as well as relationships with the other {57} agencies in this framework. Mr. Lilienthal's article (N. Y. *Times Magazine*, January 7, 1945) is devoid of any reference to this problem.[11]

It appears, indeed, that TVA has not faced the need to accept the federal structure as given, and has taken on an "all or none" attitude whose revolutionary implications for the structure of the national government are accepted, though seldom made explicit. This viewpoint is irresponsible in the nonderogatory sense of a single-minded pursuit of special goals which leaves consequences to be handled by those directly affected and by the compromises of history.

There are few organizations which can afford the luxury of a neat or complete correlation of ideas and action. The needs which lead to or reinforce the formulation of an official doctrine do not ordinarily exhaust the demands made upon an organization. To be sure, there must be some warrant in action for the general ideas propounded by the leadership, but the latter is free to select from a portion of its program a set of objectives which will reflect its primary aspirations. This fragment may become the receptacle of meaning and significance and—on the level of doctrine, of verbalization—may infuse the organization as a whole with a special outlook. But only when a leadership is reckless of the fate of its charge may it drive toward the exemplification of an administrative ideology in all phases of its activity. The TVA has been no exception. Its needs are not simple and are certainly not exhausted by those which have generated the grass-roots theory. For example:

1. In the light of its commitments to press the struggle for public power, the TVA could not permit its power program to be channeled through any of the existing agencies of the state governments in its area. Although TVA has acted primarily as a power wholesaler, it is fair to say realistically that the power program has been one of "direct action." It is true that the power distributors are local municipalities and electric power coöperatives, and these are counted as grass-roots agencies, but the difference between the operation of the electric power program and the agricultural program is too great to be designnated by the same name. In power, the TVA carries forward its own program, with stringent contractual controls. In agriculture, the TVA has shaped its program to use the existing agricultural leadership. The

power coöperatives have been built (though based on local response) with the direct and forceful intervention of the TVA itself. Such means have been ruled out in the agricultural field. {58}

2. In other fields the TVA has not always waited for the development of proper institutional channels on the local level. In the building of the river terminals, an important phase of the Authority's responsibilities in the development of navigation, local agencies were not always considered adequate. The decision to build a river terminal is made by the TVA Board, and it is difficult to determine the criteria by which the local channels are rated as adequate. Sometimes, notably in local planning, new agencies have been organized, with TVA occasionally instrumental in the initiation of new state legislation.

3. In 1942, during the wartime emergency, the TVA went ahead with the construction of Douglas Dam despite considerable protest from the area. Whatever the merits of the case for building the dam, it is evident that in this important matter opportunities for local decision were not available. While the decision to build the dam was made in conjunction with the leaders of the national government, there is no evidence that the TVA Board opposed the development. In this matter, as in many others, the TVA represents national policy and executes that policy without regard to local interests. Despite the consequent difficulties for adjustment to the area of operation, it would appear to have been consonant with the broad interests of TVA to produce the maximum possible power during wartime to bolster its position on a national scale.[12]

Beyond the fact that an official doctrine, however sincerely held, is not usually operative in all phases of an organization's activity, it is worthy of note that ideas may serve multiple needs. The multiplication of the functions it serves may strongly bolster the doctrine because of the ease with which it comes to hand as repetitive problems are faced. For it is little more than a truism that the utility of an idea—not so much in inquiry as in smoothing the path of a predetermined course of action—may easily invest it with a specious cogency, drawing attention away from objective tests and evidence. Thus the theory of "unified resource development" may (1) aid in breaking down the barriers which separate technicians within the regional authority, and (2) serve as a justification for control of resource development by a single agency. Again, "water control on the land" is an effective formula which ties general welfare objectives such as the improvement of agriculture and soil conservation to the constitutional pegs of navigation and flood control; but the concomitant idea of carrying on soil conservation in order to protect {59} the investment of the government in dams and reservoirs also serves to justify exercise of those functions by TVA rather than some other agency. A formula, if it is sufficiently pat, may be a substitute for inquiry, a possibility which is strongly reinforced if the formula also promotes a resilient self-consciousness.

UNANALYZED ABSTRACTIONS

Language used for self-protection and for exhortation develops terms which are unanalyzed, and persistently so, for their effectiveness depends upon the diversity of meanings with which they may be invested. This is well understood in the field of frankly moral injunction, emotionally formulated, designed to persuade. But the similarity between the glittering generalities of the sermon or political harangue and the apparently technical devices of administrative relationship, while significant, is far less obvious.

This similarity lies in the fact that all statements contain unanalyzed terms until and unless a specifiable context is at hand. Such ideas as "to achieve unity" or "lend assistance" assume meaning empirically only when we know how to answer the questions: with or to whom? under what circumstances? and for what? There are times when circumstances will permit only one answer, but normally a process of selection is involved which, when made explicit, provides the abstractions with a concrete reference. Usually, when methods are essentially routine, the implicit selection is understood by the participants, so that what seem to be unanalyzed abstractions have contexts adequately specified in precedent. But when formulations of procedure are elaborated into organizational doctrine, and come to have a value and function apart from technical goals, the doctrine becomes a vehicle of organizational self-consciousness, a tool for the education of the ranks, and may become progressively divorced from any definite subject-matter. In the process of being informed with an affective content, the procedural doctrine may be linked to some commonly accepted symbol — such as, in America, "democracy" — in order to share a derived and conditioned aura. Increasingly, concrete analysis will be avoided, and since unanalyzed abstractions cannot guide action, actual behavior will be determined not so much by professed ideas as by immediate exigencies and specific pressures.

It has been suggested that the Tennessee Valley Authority grew to organizational maturity under a natural pressure to come forward with a theory which would justify the very nature as well as the existence {60} of the organization. For TVA was from its inception a challenge not only to the "power trust," but among other things to the theory of government which prevailed in its area of operation: the idea of local sovereignty and "states' rights." Objectively, perhaps not always explicitly understood, it became the task of the Authority's leadership to find a means of mitigating the opposition to an organization imposed upon the area by the central government. This task was a social need, essential to the continued existence of the agency. Such a need does not, of course, mechanically determine the flow of ideas; but it creates problems and possible consequences for the life of the organization which select and reinforce convenient formulations, and which subtly reject those which carry an implicit challenge or which may tend to undermine essential loyalties.

It need not be denied that the notion of a grass-roots administration is a product of the TVA's genuine concern for democratic procedure; that concern may be associated with the fact that the specific doctrine produced has been shaped by the need to adapt the organization to its institutional environment, to come to terms, to make its peace with the existing social structure. These objectives can, in a limited sense, go hand in hand, the one fortifying the other. But the limits are soon reached. For the needs of maintaining the organization tend to drive it toward alliances and mechanisms of participation which are specific and immediately most effective. Commitments are made, traditions and habits laid down, which have inevitable restrictive pressure on the exercise of democracy. As a consequence, there is a strong tendency for the theory itself to contain unanalyzed elements, permitting covert adaptation in terms of practical necessities.

The most obvious of the unanalyzed elements in the grass-roots theory of administration is the use of such terms as "the people" and "institutions close to the people." Returning again to a formulation by Lilienthal, we may read:

> To administer national laws entirely from Washington... tends to exclude and ignore the local and State institutions the people have already set up. Usually these agencies are close to the people and understand their problems; they can often be of great value in helping to carry forward national policies. But in the nature of things an over-centralized, remote administration of Federal functions will be unable or unwilling to enlist the active partnership of these existing institutions, public and private agencies such as the universities and extension services, local and State planning commissions, State conservation boards, chambers of commerce, boards of health—the list is long and inclusive.[13] {61}

Lilienthal's theory attempts to adjust the new requirements of public policy to the traditional federal system. Without desiring or trying to reverse the trend toward increased federal authority, he aims to so exercise that authority that the existing state administrative structure may be fitted into the new pattern. In the above quotation we find specified not only the need for decentralized administration, but its avenues as well. The administration of federal authority is to use "existing" agencies "close to the people." The assumption is made that these institutions are in fact representative of the people, or that in the long run they will most closely approximate representativeness. In those circles in the Authority which defend the official viewpoint, the tendency is to speak in terms of the "long run" and to make little effort to defend the existing character of the local organizations on the basis of democratic criteria. In fact the Authority has to adapt itself not so much to the people in general as to the actually existing institutions which have the power to smooth or block its way. It therefore becomes ideologically convenient to fall in with the general practice in the area of identifying the existing agencies with the people, and permitting *de facto* leadership in the region to be its own stamp of legitimacy. Thus the grass-roots approach permits the Authority to come to terms with one of the most important elements of potential resistance by approving the maintenance and strengthening of the existing administrative apparatus at the state level. Moreover, since it may be assumed that this apparatus and its control is the primary prize in the struggle against federal encroachment, it is easy to see that the grass-roots approach permits a vital concession to the "states' rights" forces. All this, once more, is quite apart from the conscious sources of the official doctrine.

The attitude with which the TVA has had to contend may be indicated by the following quotation from President James D. Hoskins of the University of Tennessee, one of the "existing institutions" of the Valley region:

> ...as Federal aid increases—as it apparently must—what will be the outcome in the administrative field? Are we now to see the cherished democratic principle of decentralization of authority and local control drowned beneath the flood of Federal dollars? Or, is there to be such a flood? Shall the Federal government operate and control educational programs if state and local agencies cannot do so unaided? Or shall the Federal government merely provide financial aid, and leave control to the local authorities?[14] {62}

President Hoskins went on to attack new trends in federal administration resulting in the by-passing of local agencies, particularly in the field of agriculture. Yet TVA has received the public support of the governors of the states in its region, particularly on the grounds of satisfactory coöperation.[15] In this respect, that of coming to terms with key institutions in its area, the TVA has used the grass-roots approach to considerable advantage.

The grass-roots theory is not a mockery or fraud simply because it has served a function tangential to democratic values. A living organization must use tools which are at hand, with due regard to consequences for its own existence, accepting compromise as a normal state of affairs. Such formulations as "concern for the people" remain, as doctrine, essentially unanalyzed; unless certain concrete procedures are specified, such ideas are bound to be relegated to an organizational limbo, because they will not be useful in embodying the principles in action. On the other hand, once procedures are specified, then these procedures tend to become the receptacle of the positive emotional values provided by the general principle, even if what is put into practice is a theoretically unsatisfactory rendering of the original moral imperative. In action (and hence concretely), the method will be embodied in relations with some actual institutions. These will of necessity share whatever moral value may be attributed to the method; as responsive structures, they will invest the total relationship with a meaning determined by their own origins and commitments.

The authority's staff is not unaware of this problem, although there is practically no open criticism of the basic official doctrine. One longstanding member of the staff felt that the leadership had betrayed its real democratic responsibilities, and remarked:

> The grass-roots policy is merely a rationalization. It is absurd to have a federal agency trying to work through organizations which place it at a second remove from the people. There is a distinction which must be drawn between "institutional grass roots," represented by Senator McKellar, President Hoskins, and TVA's Agricultural Relations Department, and a "popular grass roots" which would be less concerned with the prerogatives of established leaders. Unfortunately, in practice, the TVA has chosen the former.

This is an extreme view, but it serves to indicate the varying content with which unanalyzed terms may be invested, and the way in which the significance of the doctrine is bound up with its concretization. {63}

Aside from the general problems of who are "the people," and the need to make explicit the criteria of selection, there are other ambiguities in the terms of the grass-roots doctrine:

1. There is a vagueness in the official doctrine about the mechanics by which local influence may be brought to bear upon a decentralized agency. It is well known that, where administrative control from the center is weak, the field officers of an organization may tend to become representatives of their area. As such, they are subject to the informal influence of close association with local problems, including the pressure of a local public which must be dealt with on a day-to-day basis. An independent agency may indeed identify itself with the welfare of its region, and shape its techniques of administration in local terms. But it is not clear that the people of the Tennessee Valley have been afforded any significant control over the operations of the TVA beyond the normal channels of congressional appeal. The responsibility for decisions of serious consequence lies with the Congress. The Tennessee Valley, as a region, does not make decisions on such important matters as the construction of dams or the expenditures of the regional authority.

2. It may be questioned whether the location of the central office of an agency is decisive for the quality of being "close to the people." Long association, involving acquaintance with field personnel as well as with the habits, customs, and outlook of an agency, may be of equal importance. This is especially true if "the people" is identified with its organized expressions, existing local governments and private associations. These institutions may build up a stable set of relationships with agencies controlled from afar, and yet feel that they are close. The defense of the Bureau of Reclamation and the Corps of Engineers by many local groups from the Missouri Valley, and their fear of some new

agency which, though decentralized to the area, might be alien, is a good example of such a relation. The issue here is not the defense of one approach against the other, but simply to point out that the meaning of "close to the people" is unanalyzed.

3. The significance of "working through established agencies" may vary, depending, for example, on the relative strength of the local organizations. There may be effective coöperation among equals, or dominance by one side or the other. Much may depend on the extent of discretion and the degree of common agreement on general approach. The development of administrative constituencies may significantly shape the concrete meaning of the method.[16] {64}

4. "Coördination" is another term whose meaning will be derived from the procedures which are established for its effectuation. Like "unity," this is a word which is often used vaguely in order to avoid making real objectives explicit. It is an honorific which must grace all administrative programs. But coördination involves control; and it is precisely the concrete lines of authority implied in any given system of coördination which constitute its meaning. Coördination may range from the mere circularization of ideas through the definition of jurisdiction all the way to the most thoroughgoing *Gleichschaltung*. In the TVA, one leading official has made a distinction between the "coördination of agencies" as the function of a regional authority and the "coördination of programs," with the latter considered as proper. But of course if this involves the elimination of other agencies, whose programs are taken over by the regional authority, it will not be highly valued, as coördination, by the agencies in question.

5. "Participation of the people" is rated highly; but here again its significance depends upon such factors as the meaning of the program to the individuals concerned, the level of education, the actual locus of decision, and the nature of the organizations through which participation is achieved. Participation may range from mere involvement by means of devices established and controlled by the administrative apparatus to spontaneous organization based on and integrated with existing community patterns of coöperation.[17]

ADMINISTRATIVE DISCRETION

The propaganda of the "outs," of those who need not take responsibility for a definite course of action, may retain its unanalyzed elements with impunity; but those who govern (in organizations large or small) are forced to act, and to assume responsibility for the consequences of events. As a leadership exercises its discretion in action, it generates a history which serves to make concrete the unanalyzed abstractions of its doctrine.

Discretion is a process of selection—above all the selection of tools for the execution of policy. But these tools are social facts with lives of their own, with needs and demands, with determined responses which do not permit an easy subordination to alien or indifferent ends. Tools are recalcitrant; they ask a price, the price of commitment. He who would induce coöperation must agree to be shaped in turn, to submit to pressure upon policy and action. This give and take may not appear {65} in its most significant forms in a contract or agreement. It may never be verbalized.

The influence of recalcitrant tools upon policy may not be readily apparent. It may not be directed to the specific program which prompted coöperation. Since the consequences of coöperation ramify widely, a coöperating group may execute a given policy initiated by another without substantial deviation, but at the same time it may make demands in another field as the price of continued coöperation. The meaning of the relationship of author and agent, of initiator and tool, will not be understood until the consequences of coöperation for decision making can be traced, including the effect upon the position of the organization among its contemporaries, and upon its nature in the light of long-run objectives.

Management strives for discretion; a leadership seeks freedom of movement, a range of choice, so that it may more ably invest its day-to-day behavior with a long-run meaning. Prestige and survival are not normally accepted as legitimate ends of administrative behavior, especially in government and among the subordinate units of a larger organization. But prestige and survival, even for the smallest unit (so long as there are those who have a stake in it), are real factors in decision, as all participants know. Lacking formal channels, necessarily concerned with more than a substantive program, management requires discretion as a means of introducing the factors of prestige and survival into its choices among alternative methods for the execution of formal policy. This factor is reinforced as the role of the managerial expert grows in importance. Professionalization in management, providing ideals of method and loyalty to managerial objectives, creates a "responsible management." But, as the price of responsibility, the expert insists upon an enlargement of discretionary powers. He joins the ranks of other experts in denying to the layman the right to judge among alternatives when these lie within the province of specialized experience.

Discretion is selection; and it is also intervention. It enters upon an existing social situation, creating disequilibrium and anxiety until some new adjustment restores order and security. Intervention involves consequences for (1) enduring factors always present where prior social organization exists; and (2) conditions of imbalance or conflict which happen to exist at the time of intervention.

1. Administrative intervention cannot escape the force of precedent. Normally, the attempt to institute a program will run up against a {66} network of prerogative and privilege, which will have to be dealt with in some manner before operations proceed very far. Except in extreme cases, these interests will not be liquidated or even absorbed, but will remain as centers of power which the organization

must take into account. This situation soon creates the problem, among others, of the representation of interests. This may take explicit form,[18] or, when the real pressures are strong and no formal avenues are permitted, representation may be informal and sometimes covert. Indeed, where the selective process implicit in discretion may invite criticism, the process itself may become informal and covert. When there is a serious uneasiness about informal representation, an ideology may be developed, or an existing doctrine, perhaps otherwise neutral, may be reinforced. A doctrine of legitimacy may be elaborated, justifying a specific selection of representatives when alternative possibilities are present. Where a program is broad, and representation of a general interest group or class is at issue, choice by the agency often signifies intervention among the competitors for leadership within the client group itself. This is a decisive act of discretion, and may often commit an organization to a faction with which it has no sympathy and to a conflict it does not desire; or such discretion, when its consequences are recognized, may be a means of modifying the social structure with which the organization has to deal.

2. Discretion may occur in a context of contending social forces. The loyalties of an administrative leader may be weaned away from exclusive attention to the organization with which he is identified when the interests of a social class or group are at stake in a general struggle. In such a context, administrative decisions may be made in terms of consequences for the over-all contest, with the needs of the particular organization subordinated. An extreme case is the role of Communists in labor unions, in particular their tendency to subordinate the interests of a union to the needs of the general defense and extension of the policies of the regime in Soviet Russia. But the pattern is more general: in a period of anticlerical struggle, it will be expected that a devout Catholic administrator will devote some attention to the consequences of his actions for the welfare of the Church; in a period of considerable ethnic tension, the decisions of a Negro or a Jew might be expected to be shaped by considerations tangential to the special program of his organization; when sectional feeling runs high, an administrator in a {67} federal agency might subordinate his loyalty to the agency as such to the struggle for the advancement of sectional goals.

The problem of administrative discretion is customarily raised in terms of the potentially arbitrary action of quasi-legislative or quasi-judicial regulatory bodies.[19] This is of great importance as the practice of delegating authority to the executive power is extended. But from the point of view of organizational analysis it is the internal relevance of discretion which is especially significant. Administrators commonly develop commitments to methods and to specific organizations with which convenient relationships have been established; these in turn are related to client groups with ends and commitments of their own. As a consequence, a system of mutual dependence and aspiration develops which cuts across organizational lines. Hence the exercise of discretion must be analyzed with reference to the structure of the governmental system as a whole, as well as to the evolution of the particular organization itself. The execution of a program may be materially affected, in the long run, by the managerial structures which are built and sustained in connection with it. Insofar as discretion is permitted in the building of any given organization and those related to it, a measure of control over policy itself will in effect have been delegated.

The TVA's grass-roots doctrine seeks to maximize discretion. Depending upon the objectives of public management, this may be an unimpeachable program. It is not the concern of this essay to offer an appraisal. What is significant here is that the exercise of discretion in a concrete social context and under the pressure of organizational needs created a problem of interest representation, emphasized by action within a situation marked by controversy among groups and agencies.[20] In addition, the elaboration of a doctrine justifying the maximization of discretion poses for the federal structure significant alternatives unanticipated by the founders of the Authority. The wide measure of corporate freedom afforded the TVA was related to its business function and justified by the experimental character of the project, as well as by the exigencies of the public power controversy. This has been extended by the TVA itself to a program for the decentralized administration of regular federal functions. On this view, the straightforward alternatives implicitly posed by the Authority for {68} the future of the federal government are (1) no independent regional agency; or (2) the subordination of departmental prerogatives to the discretion of the regional agency. Since it is unlikely that the federal departments or their organized clienteles will willingly abandon structures and precedents established after considerable effort, it is possible that the concrete interpretation of grass-roots administration, evidenced in the discretion exercised by TVA, may endanger the general idea of decentralization itself: some of the older established agencies have interpreted their experience with TVA to mean that an independent regional agency cannot effectively coördinate federal agencies in the field. The TVA leadership feels that it need not be concerned about the consequences of its version of decentralization for the federal departments, for, it contends, a democratic perspective is concerned not with the prerogatives of officials but with the needs and desires of the people. But this view breaks down as we note that the special relation of these federal agencies to client publics who have a stake in their continued existence makes of them something more than simply bureaucratic domains. Hence the problem of discretion assumes its full significance only as we see its consequences for organizational character and relationship, internal and external.[21]

The organizational consequences of discretion are not often made explicit. However, one TVA official, discussing the future of the Authority's financing, made the following statement:

The most obvious result of comprehensive revenue-financing power is to give the Authority largely independent discretion in the expenditure of funds on "development activities." The miscellaneous activities falling within this category are the ones spoken of in the statute with the least clarity. At the same time, they constitute the integrating cement of the Authority's "regionalism." Also, they are the activities most subject to conflict with the various activities of the various line departments of the federal government, and of the Valley states. The content of this development activities program has expanded and changed markedly in conjunction with the evolution of the Authority's concept of an integrated regionalism. With the completion of the major construction phase, the real character of the Authority will depend upon the character of its development activities program.[22]

The TVA does not have and has not sought the measure of discretion involved here, but it is clear that a consideration of the implications of discretion raises problems significantly related to the emergent character of the organization itself. {69}

INHERENT DILEMMAS

Tension and dilemma are normal and anticipated corollaries of the attempt to control human institutions in the light of an abstract doctrine. Social structures are precipitants of behavior undertaken in many directions and for many purposes. Mutual adaptation establishes only an uneasy equilibrium. This in turn is continuously modified and disturbed as the consequences of action ramify in unanticipated ways. Practical leadership cannot long ignore the resistance of social structure, and is often moved thereby to abandon concern for abstract goals or ideals—for which it is often criticized out of hand by the moralists and idealists who lack experience with the vicissitudes of practical action.[23] But a leadership which, for whatever reason, elects to be identified with a doctrine and professes to use it in action, is continuously faced with tensions between the idea and the act. Ideological symbols may fulfill useful functions of communication and defense and may be long sustained as meaningful even when effective criteria of judgment remain lacking; but an act entails responsibility, establishing alliances and commitments which demand attention and deference.

This is not to suggest that ideals are futile and abstractions useless. Tension does not mean defeat, nor does dilemma enforce paralysis. It is precisely the problem of leadership to find a means, through compromise, restraint, and persuasion, to resolve tensions and escape dilemmas. But in doing so, attention must be directed to the real forces and tendencies which underlie its difficulties. This is the constructive function of analysis which seeks to take account of structural rigidities and the indirect consequences of executive action. "Where such analysis is considered destructive, it is usually because doctrine, assuming an ideological role, is not meant to be analyzed. In extreme cases, un-analyzed doctrine ceases to operate in action at all, and the real criteria of decision are hidden in a shadowland of unrecognized discretion determined opportunistically by immediate exigency.

The TVA, in relation to its policy of grass-roots administration, is not immune to such difficulties. Though seldom made explicit, sources of tension are recognized by members of the staff, and have already entered into the process of administrative decision. Among these may be noted:

1. *There is a dilemma of doctrine and commitment, or of the abstract {70} and concrete.*[24] This dilemma is the most general source of the tensions inherent in the grass-roots approach, for it inheres as well in all behavior which involves both the verbalization of ideas and a set of specific activities. Doctrine, being abstract, is judiciously selective, and may be qualified at will in discourse, subject only to the restrictions of sense and logic. But action is concrete, generating consequences which define a sphere of interest and responsibility, together with a corresponding chain of commitments. Fundamentally, the discrepancy between doctrine and commitment arises from the essential distinction between the interrelation of ideas and the interaction of phenomena. The former is involved in doctrine, the latter in action. Of course this is the ground of the normal necessity to revise decisions and even over-all doctrine in the light of events. The tension between the abstract and the concrete is resolved through continuous executive action. However, where doctrine itself creates commitment, as in the institutionalization of policy, executive decision is not readily reversible. Policy which ostensibly should be determined on the basis of a scientific appraisal of practical means for the achievement of formal ends becomes invested with prestige and survival value and may persist as official doctrine despite a weakening of its instrumental power. Whatever instrumental capacity the policy does have is related to informal rather than professed goals.

2. *There is also the dilemma of consent and conformance,*[25] *or of selective decision and total involvement.* Democracy as method, and the grass-roots policy as method represent processes of decision. Decision, however, demands only the partial consent of the participants, who are involved only obliquely in their capacity as voters or choosers. But the execution of decision is a matter of action, which tends to involve the participants as wholes. Hence coöperative action, as Barnard says, "requires substantially complete conformance."

In other words, while the choice of a given course of action may be conceived of as involving the individual or group only to a limited degree, in fact there is a tendency for circumstances to demand more extensive involvement. Organizational action, once initiated, tends to push onward, so that the initiator may be enmeshed in new relationships and demands beyond his original intention. Here again the key word is "commitment." Every executive knows that the initiation of a new course of action is a serious matter precisely because of the risk involved that the establishment of precedents, of new machinery and {71} new relationships, the generation of new and complex interests may make greater demands upon his organization than he can presently foresee. The problem is not one of inner impulse, but rather of the structural forces which summon action and constrain decision. This structurally induced tendency is roughly comparable to the notion of adience as applied in the theory of animal drive.

Conformance, indispensable to the completion of the movement begun by decision, has therefore a qualifying effect and creates an inescapable tension. This means, in respect to the TVA, that while its decentralist policy may be instituted for special reasons (referrable in part to the ideals of the leadership as well as to other factors), nevertheless there will be a tendency for the organization as a whole to be shaped by the process of conformance. It will have to go farther in carrying out the policy than it may have originally intended; the formal process of consent will have been transformed into concrete institutional relationships, generating new and unlooked-for demands. Generally, the problem of checking conformance is present in all administrative decisions concerning relationships established under the grass-roots formula.

3. *So far as discretion is delegated, bifurcation of policy and administration is reinforced.* It is basic policy inside the Authority that planning and execution should be united in a single administrative organ. But the delegation of functions involved in the grass-roots approach makes insistence upon this principle of unity somewhat anomalous. The channeling of programs through independent agencies necessarily delegates discretion, hence programs may be extensively modified in execution. The dilemma is only made more explicit if controls are instituted which operate objectively to transform the independent local agency into an administrative arm of the Authority: execution may be brought into line with policy, but the grass-roots objective will have been undermined. The dilemma is mitigated to the extent that a denial is made of the possibility of a difference in objective between the initiating and the executing agency. An attempt in this direction made in the TVA's agricultural program will be described in a later chapter.

4. *Theories about government become preëmpted by social forces.* It is unrealistic to estimate the implications of a theory of government on the basis of its abstract formulation. The propagation of a point of view carries with it an often unwished-for alliance with others who, for their own reasons, are espousing the same or a convergent doctrine. A theory about method may at any given time be linked with a special substantive doctrine; support for the method many imply a position on {72} program, and conversely. Thus TVA finds itself in the camp of the supporters of "states' rights" both in its criticism of over-centralization in government and in its support of the state governments—as "regional resources"—in its area of operation. But at the same time, in respect to the complex of political issues summed up in the extension of "posi-

tive government," the Authority's point of view is very far from that of the general run of supporters of local sovereignty. TVA is therefore in the continuously ambivalent position of choosing between an emphasis on method and an emphasis on substantive program, or more accurately, of assigning a priority to one or the other. To the extent that the movement for planning and a strong federal government oriented along welfare lines is identified with the existing federal structure, the TVA leadership tends to emphasize method as basic and cut itself off from the general welfare movement.[26] But it cannot divorce itself completely from its antecedents, and the ambivalence persists.

5. *Emphasis on existing institutions as democratic instruments may wed the agency to the status quo.* A procedure which channels the administration of a program through established local institutions, governmental or private, tends to reinforce the legitimacy of the existing leadership. This is especially true when a settled pattern claims the exclusive attention of the agency, so that other groups striving for leadership may find their position relatively weakened after the new relationships have been defined. In strengthening the land-grant colleges in its area, the TVA has bolstered the position of the existing farm leadership. There is some evidence that in the process of establishing its pattern of coöperation, TVA refrained from strengthening independent colleges in the area not associated with the land-grant college system. Again, the relatively dominant role of the American Federation of Labor unions in TVA labor relations, especially as constituting the Tennessee Valley Trades and Labor Council, is objectively a hindrance to the development of labor groups having other affiliations. In general, to the extent that the agency selects one set of institutions within a given field as the group through which it will work, the possibility of freezing existing social relationships is enhanced. At least in its agricultural program, TVA has chosen to limit its coöperative relationships to a special group, so that the potential or inherent dilemma has been made explicit. {73}

6. *Decision at the grass roots may be inhibited by the system of national pressure groups.* When important issues become crystallized in the program of a group organized on a broad scale with a national leadership, local decisions may be influenced primarily by their effect on the outcome of over-all controversy. The local problem is appraised not for its own sake but for the influence of a local decision on the general bargaining position of the leadership. Thus a local branch of the American Farm Bureau Federation or of a CIO union may have its attitudes framed for it by a long-run national strategy, as by a pending bill in Congress for the transfer of functions from one agency to another, or an impending organizing drive, or a national election. To the extent that a decentralized governmental agency is influenced by decisions made by local groups in national terms, it would appear that the grass-roots approach becomes the victim of the growing centralization of political decision.

7. *Commitment to existing agencies may shape and inhibit policy in unanticipated ways.* When the channels of action are restricted, programs may be elaborated only within the limits established by the nature of the coöperating organizations. The traditions and outlook of an established institution will resist goals which appear to be alien, and the initiating agency will tend to avoid difficulties by restricting its own proposals to those which can be feasibly carried out by the grass-roots organization. Where the grass-roots method is ignored, new institutions may be built, shaped *ab initio* in terms of the desired program. An attempt to carry forward a policy of nondiscrimination (as against Negroes) will not proceed very far when the instrument for carrying out this policy—usually as an adjunct of some broader program—has traditions of its own of a contrary bent. Moreover, the grass-roots policy voluntarily creates nucleuses of power which may be used for the furtherance of interests outside the system of coöperation originally established. Thus the TVA distributes electric power through electric power boards which are creatures of municipalities, with the contractual reservation that surplus income shall be used only for improvements in the system or for the reduction of rates. But the question has been raised: what if pressure arises to use surpluses for general purposes, that is, to finance nonpower functions of the municipal governments? And what if the state governments undertake to tax these surpluses, because of a restricted tax base and unwillingness to institute a state income tax? The logic of the grass-roots policy might force the Authority to agree. However, it is perhaps more likely that the Authority's {74} commitment to function as a successful power project would take precedence over the grass-roots method.

8. *Existing agencies inhibit a direct approach to the local citizenry.* The participation of local people always takes place through some organizational mechanism, notably voluntary associations established to involve a public in some measure of decision at the end-point of operation. But such associations are commonly adjuncts of an administrative agency which jealously guards all approaches to its clientele. If, therefore, a federal agency establishes coöperative relations with such an agency, it will be committed as well to the system of voluntary associations which has been established. Hence the channels of participation of local people in the federal program will be shaped by the intermediary agency. In respect to its closeness to the people, the status of the federal government may not, in such circumstances, be materially altered. Viewed from this perspective, the grass-roots method becomes an effective means whereby an intrenched bureaucracy protects its clientele, and also itself, from the encroachments of the federal government.

IMPLICATIONS FOR U. S.–TVA RELATIONS

The grass-roots theory has served the special needs of the TVA as an organization, above all the need for acceptance in its area of operation.

It is not surprising, therefore, that the idea has not been wholeheart-
edly welcomed by agencies of the federal government whose responsibilities are national rather than regional; and that the TVA has chosen
to clash head on with some Washington departments, thus failing to
make an adjustment to that part of its environment constituted by the
existing structure of the federal government. In the early days of TVA,
the so-called "New Deal" agencies were unwilling to interfere with a
social experiment that was an important symbol in the over-all political
struggle. But as TVA's position became secure, such organizations as
the U. S. Department of Interior began to consider the implications of
the idea of an independent regional authority for their own future.
Doubts were transformed into urgencies as the movement for the
extension of the TVA approach to other regions gained headway.

The extension of the TVA administrative structure, especially its
managerial autonomy, to other areas, poses far-reaching alternatives
for the future of the federal government. If broad responsibilities for
regional development are lodged in a single agency, and such agencies
are multiplied so that they include virtually the entire area of the
nation, the operational functions of the developmental agencies in {75}
Washington may be substantially curtailed or even eliminated. Driven
to a logical (if somewhat farfetched) conclusion, such agencies would
primarily assume staff functions, involving a radical bifurcation of
planning and execution. At the other extreme, regional agencies would
themselves be purely advisory bodies or operating arms under the control of a federal department, or would not exist at all. These long-run
implications have not been lost to the leading participants in Washington and Knoxville.

For the TVA the question of the structure of future regional authorities is not academic. Partly it involves the prestige of the leadership,
because of personal and professional commitment to an embattled
method.[27] But perhaps more urgent is the realization that TVA itself
may not long maintain its autonomy if a national decision is made
against the autonomy of other regional agencies. The Authority would
then become an anomaly, and in time would be coördinated with the
structure as a whole.

"If there is one thing," Lilienthal has said, "which the nine years of
TVA experience have proved to us, it is this: unity of management
located physically in the region is a *sine qua non* in getting results."
With this conviction as a foundation, he has argued repeatedly that the
establishment of regional development authorities cannot be viewed
apart from the question of administrative structure, insisting that the
TVA pattern of autonomy is essential to successful operation. This
autonomy includes such minimum essentials as freedom from veto by
such control agencies as the General Accounting Office; freedom from
control by the U. S. Civil Service Commission; freedom to use current
revenues for operating purposes; and unified control over its own developmental programs so that the policy of coöperative administration

with and through local and state agencies may be effectuated. The following informal statement by Lilienthal indicates the quality of urgency with which the debate (at least from the TVA side) is suffused:

> We have observed a tendency among those who wrestle with this problem (of developing natural resources on a regional basis) to assume that administrative arrangements in dealing with regional development problems are relatively unimportant. I cannot emphasize too strongly that our experience shows that quite the contrary is true. The things that have happened in the Tennessee Valley, for which the TVA is in any way responsible, would not have happened but for the fact that there is nothing vague, or "fuzzy," or indeterminate about the central administrative responsibility {76} and commensurate authority and autonomy which the TVA carries under the law. Coöperative arrangements, devices of coördination, and the fullest collaboration among federal and state agencies are characteristic of the TVA's methods of getting the job done, but these methods would go for naught, in our judgment, unless there were a central management able to make important decisions and to provide a core of leadership in the region, reasonable autonomy, and independence from minute direction or delays of paralytic indecision from Washington.

The issue, Lilienthal suggests, may be difficult to resolve if the prerogatives of the Washington bureaus must be taken into account, but it is simply answered if "the people, the public, have to be satisfied." But just these prerogatives are of vital importance to the existing federal agencies: hence the conclusion that TVA has elected not to make the same adjustment to the existing pattern of national government as it has to the existing pattern of state and local government. The grassroots doctrine is a vehicle for the accomplishment of the latter; at the same time it is an instrument of disaffection from the Washington establishments. It is, of course, not the purpose of this discussion to enter the debate in question, or to make any appraisal of relative merits. We are interested here only in the question of the implications of the grass-roots doctrine for the position of the TVA as an organization.

The self-consciousness of administrative agencies is a factor in the behavior of all leading participants. A formulated doctrine is a prop to self-consciousness and a vehicle for its expression. This assumption is important for the interpretation of the grass-roots theory as a means of defining the approach of TVA to other organizations. The role of the doctrine becomes significant as it informs the handling of current issues, as it invests them with a long-run meaning. This is true of all organizations, though some are more prone to verbalize than others. An example in the TVA, which highlights the potential strains in its relations with a federal agency, may serve to make this clear. In 1941, the Department of Agriculture established certain quotas for increased agricultural production, channeling its program through the State

USDA Defense Boards and State Land-Use Planning Committees, which included representatives of the state extension services of the land-grant colleges. This procedure was followed in the Tennessee Valley without clearance through the Correlating Committee jointly representing the TVA, the USDA, and the Valley states land-grant colleges. Presumably, the Department felt that its responsibilities under the defense program could best be carried out through organizations which it, independently of relations to the TVA, had established on the state and local level. The Department's program for Tennessee called for the expansion of quotas {77} for feed corn, which involved what it considered a small net increase in clean-tilled acreage. The TVA took issue with the Department on this matter, suggesting that the desired increase in livestock and livestock products could better be obtained by expanding the acreage, production, and yield of pastures and meadows, thus continuing the conservation policies, supported by the Authority, of reducing row crops and aiding in the development of cover crops by the distribution of phosphatic fertilizers. The Department maintained its viewpoint, insisting that it was coördinating con-servation farming with the national interest in increased production through the agency of state agricultural leaders who had included considerations of erosion and related conservation factors in their determination of local plans and quotas.

This difference in approach seemed to the TVA a significant example of inadequate definition of national policy in regional terms. The Authority, of course, could do nothing about it, but its interpretation of the incident in self-conscious and long-range terms is significant. One leading TVA official pointed out that he felt there was a real danger that the Department, in expanding food production, might undertake activities inconsistent with the program in the Valley, which integrated crop production with the regional problem of land and water control. Moreover, he pointed out, "the current trend of deciding all such matters at Washington, without the careful consideration of special local circumstances which the TVA and coöperating agricultural colleges can supply, *is a threat to the long-time effectiveness of regional organizations such as TVA.*" (Emphasis supplied.) Thus, the Department's involvement of state and local agricultural committees, presumably a grass-roots approach, is discounted, and the implications for TVA as the integrating regional agency has come to the fore. This official stressed the significance of the issue for the long-time destiny of TVA, and pointed out that in the postemergency programs TVA should "fight to prevent the entire regional approach to national problems from being submerged by the pressure of emergency needs, and by the dangerous tendencies toward overcentralization of public policy determined in Washington." In this way, the grass-roots doctrine, when identified with the indispensability of TVA, serves to give meaning to current problems, so that they are related to long-time organizational strategy. It is interesting that the TVA program is referred to as the

"Tennessee Valley region's program of land and water resource development" involving a somewhat gratuitous identification of the TVA, as an agency, with the region itself. The USDA program, on the other hand, is considered {78} as divorced from the region, despite its coöperation with many of the same local organizations with which the TVA works. In this case, with respect to federal agencies, the grass-roots doctrine functions primarily to defend the TVA's autonomy and its central status with respect to the region.

The point of view of the Washington agencies, rejecting the idea of autonomy for regional authorities, has been made clear as proposals for the extension of the TVA idea have come up for consideration. Several of these points of resistance to the TVA approach on the part of federal agencies may be adduced:

1. In 1941 a bill was introduced in Congress which proposed to include the Cumberland River and its basin within the provisions of the TVA Act of 1933, thus widening the jurisdiction of the Authority to a contiguous river basin. This measure was opposed by the Department of Agriculture, which prepared a report outlining its views for the Commerce Committee of the U. S. Senate. It was pointed out that the TVA has been made responsible in its region for activities within the field of agriculture and conservation for which the Department of Agriculture had a nation-wide responsibility. Because of the TVA's program, the Department had not extended its erosion control work to the Tennessee Valley. However, even though satisfactory arrangements had been made with the TVA to avoid duplication, it was felt that the principle involved was not sound and should not be extended. The TVA arrangement, the Department held, was tolerable as an exception, but could not be accepted as a rule. The Department expressed its appreciation of the need for planning in the development and conservation of natural resources, and favored the establishment of regional planning agencies which, however, would have authority only to recommend projects to other agencies, Congress, and the President, but would not have authority to carry them out on their own account. Thus the TVA approach was rejected in principle, posing for regional agencies the same logical alternative which TVA implicitly posed for the federal departments—a staff or advisory status.

2. In a preliminary report on the Arkansas Valley, issued in 1942 by the National Resources Planning Board, including a comprehensive economic and social plan for the region, the administrative recommendations failed to follow the TVA pattern. Among the 45 persons listed as having made material contributions to the preparation of the plan there were:

7 from the Bureau of Agricultural Economics, USDA {79}
4 from the Soil Conservation Service, USDA
4 from the Farm Security Administration, USDA
2 from the Forest Service, USDA
7 from the National Resources Planning Board

2 from the Oklahoma College of Agriculture
2 from the University of Arkansas
1 from the University of Texas
1 from the Southwest Coöperatives
2 from the Rural Electrification Administration

The Reclamation Service, Corps of Engineers, and other federal bureaus were actually to execute the program. It is not surprising that these agency-minded men did not vote a lack of confidence in their own organizations. But it is significant that the NRPB's planning was (and had to be) influenced by the existing departmental structure. Even if the voice of TVA were raised in its councils, it could scarcely be expected to carry much weight. In the minds of the TVA officials, the NRPB was, as one of them put it, "preparing to butcher the TVA idea" in its report. In any case, it was clear that even an organization like NRPB, in the Executive Office of the President and separate from the operating departments, could not escape the federal vortex. This does not mean that the NRPB as an organization was opposed to the idea of regional decentralization of federal functions, or even to the authority method as such. It does mean that a federal planning agency with national responsibilities, leaning upon existing federal agencies for coöperation, could not embrace a doctrine which might be interpreted as basically inimical to the long-time interests of established organizations within the federal structure. *The TVA leadership does not appear to have used the same criteria of realism in approaching the possible response of the federal bureaus as it has in taking account of the interests and prerogatives of state and local agencies. This discrepancy seems to provide a warrant for the conclusion that TVA found adjustment to its area of operation a prior necessity; in making that adjustment, it committed itself to a theory which could not long be maintained without tension and conflict.*

3. The undercurrent of debate and disaffection which for some years had characterized TVA's relations with a portion of the federal structure became most clearly evident in 1945, in the course of public hearings on a proposed Missouri Valley Authority.[28] The problem of administrative {80} method was clearly highlighted because an extensive program of flood control, irrigation, and power development had already been authorized by Congress, placing responsibility for execution in the Corps of Engineers and the Bureau of Reclamation. In respect to this program, therefore, the problem was not what was to be done so much as by what means to do it. The Bill under discussion provided for an authority which might carry out the water control program on its own account, as well as make plans and initiate projects for the economic and social development of the region. The MVA was to be modeled after the TVA in its structure, with the same basic autonomy.

Among government officials, the chief antagonist of an autonomous MVA was Harold L. Ickes, then Secretary of the Interior.

While he did not oppose the idea of regional authorities responsible for over-all conservation and development, Ickes wished to obviate the threat to the existing departmental structure implicit in the extension of the TVA pattern. His objections appeared to be based on the experience of the Department of Interior with the TVA, as indicated in the following statement:

> . . .almost without exception, the assistance rendered TVA by the bureaus of Interior, and also, I might add, by bureaus of the Department of Agriculture, was not always accepted by TVA happily. At times, there has seemed to be in TVA a feeling that it was all-seeing, all-knowing, and all-providing within its valley, and that other government agencies were interlopers, even when they sought no more than to provide customary Federal services; at other times unfortunate ventures in untried fields resulted in the acceptance by TVA of proferred assistance, notwithstanding prior rejections. Fortunately, on the whole, that hard taskmaster—experience—seems to have convinced TVA, after it gradually lost the first exuberance of youth, that success could sometimes better and more economically be achieved through existing agencies of government. Yet, even today, there are some fields of endeavor where needless and expensive overlappings exist between TVA and other governmental agencies.
>
> It is not too much to say that one of the most important lessons which we have learned from the TVA experiment is that time, money, and manpower can be saved if future authority legislation incorporates some element of compulsion to insure efficient collaboration between authorities and the rest of the Federal government.[29]

Ickes deprecated the idea that TVA's autonomy was the primary element in its achievements, stressing the aid provided by the Public Works Administration and the Bureau of Reclamation in the early days of the Authority. The former arranged for considerable financial support during a period when the constitutionality (and hence the very existence) of TVA hung in the balance. "It was this support," the Secretary testified, {81} "that made it possible for TVA to go forward with its power program and become what it is today." Yet despite the claim of TVA that it coöperates fully with federal agencies, Ickes insisted that the TVA pattern in other regions would create a "constant pressure for exclusion" of older organizations.

For this analysis, the significance of Ickes's attack on TVA (its merits are not here considered) is that TVA's insistence upon autonomy, buttressed by the general doctrine of grass-roots administration, militated against an effective adjustment to the federal administrative structure. Thus TVA's insistence[30] upon its own terms if the authority method is to be extended, tends to bolster political opposition to all regional development agencies, autonomous or otherwise. This becomes significant as it is recalled that, in the context of national controversy over social issues, TVA remains a political symbol.

The MVA debate emphasized what appears to be the anomalous position of TVA in its insistence on local autonomy as essential to the authority method. Whatever its own desires, TVA cannot help being caught up in the general struggle over the extension of positive government. In that struggle, some federal agencies, especially those inaugurated or expanded under the New Deal, are pitted against forces which oppose planning and paternalism. The latter tend to be associated with states' rights and antibureaucratic slogans, with which the TVA grass-roots doctrine seems to converge. But that convergence cannot erase the fact that the primary meaning of TVA as a symbol is not its administrative structure but its role as a "microcosm of sovereignty," with the broadest powers of the federal government for resource development and the extension of the general welfare. The TVA idea, in the eyes of its leadership may more and more come to be identified with a method of administration, but where issues are vital no administrative sleight-of-hand will serve to obscure the real meaning of the social symbol. This was made clear by Will M. Whittington, U. S. Representative from Mississippi, among others, in testimony at the MVA hearings. Whittington asserted his opposition to the authority or regional plan, and began his list of reasons with the following:

> 1. Authorities are advocated by the spenders and by those advocating projects unable to stand on their own merits. They are advocated by the planners and by the planning agencies. They are advocated by those who believe in reforming and remaking America. They are advocated by those who believe that the Government should do for the citizens what the citizens should do for themselves. There is a place for {82} public improvements. But there is a place for private initiative. There is no place for regimentation.

> 2. Authorities contemplate uncontrolled bureaucracy that leads to irresponsible bureaucracy. Authorities are the dream of bureaucrats. They involve the establishment of a super-agency responsible to no one but itself.[31]

With the issue defined in these terms, the protagonists of the TVA method might be expected to subordinate questions of administrative structure to unity on the central points in controversy, the concepts of positive government and regional planning. But it is likely that the TVA leadership, or any similar leadership, would find small comfort in a victory for the principle of regional planning at the cost of its own administrative autonomy. At the same time, the proponents of that principle outside the government tend to lend their support to whatever specific measure is currently an issue, if it will advance political frontiers. Neither TVA nor the other interested federal agencies will make decisions in purely political terms, for unlike the nongovernmental groups, they have a stake in the consequences of specific measures for particular administrative organizations. Structures embody interests, which demand that leadership take heed of present doubts and future threats. In this way, the administrative leaders may well be disqualified from participation in a single-minded pursuit of political goals.

---

[1] For those to whom the given organization is a vital instrument for personal satisfaction, the outcome is virtually inevitable. One division chief pointed out that, in the TVA, the group of younger men whose careers were bound up with the fate of the Authority was far more ready to adapt itself to new policies framed in terms of "realistic" adjustment than was the earlier leadership, the reputations of whose members were already established.

[2] Donald Davidson, "Political Regionalism and Administrative Regionalism," *Annals of American Academy of Political and Social Science,* 207 (January, 1940), 141.

[3] L. A. Chapin in Denver *Post,* May 3, 1946. This article was part of a series entitled "Denver Post Survey of TVA."

[4] See Donald C. Stone, "Notes on the Governmental Executive," *New Horizons in Public Administration* (University, Ala.: University of Alabama Press, 1945), p. 58.

[5] Of course, the elaboration of a doctrine is not the only avenue to, or warrant of, solidarity in formal organizations. Personal loyalties, habit and custom, as well as the tendency of organizations to function as alter egos, are among the many sources of informal solidarity. It should also be pointed out that this organic quality has other consequences not always considered useful, among them a rigidity which makes it difficult to change or elaborate the formal purposes of the organization. Hence new functions, though in the same program area, may require new organizations. Nevertheless, no administrative leadership may overlook the values for internal communication of developing a unified outlook on the part of key personnel. See C. I. Barnard, *The Functions of the Executive* (Cambridge: Harvard University Press, 1938), p. 87: "The inculcation of belief in the real existence of a common purpose is an essential executive function. It explains much educational and so-called morale work in political, industrial, and religious organizations that is so often otherwise inexplicable."

[6] The analysis of origins, while relevant here, is not crucial Any ideas maintained and elaborated by the leadership—whatever their origins—may fruitfully be subjected to functional analysis. However, the significance of such an analysis is strengthened if the origin as well as the meaning of the doctrine is referrable to the needs of the organization. This does not mean that the motives of its espousers are called into question: we can tell very

little about them, and it is fair to assume that Lilienthal and those around him were activated quite sincerely by a deep concern for democracy as a moral problem. Let it be noted that our entire analysis is rooted in an interest in the meaning of events, stressing such terms as "function" and "consequence." This is quite aside from the consciousness or moral quality of the participants.

[7] House Doc. 15, 73d Cong., 1st sess. (1933).

[8] As formulated at an administrative conference, Knoxville, July 14, 1943.

[9] See *Hearings* on S. 2555 (Conservation Authorities Bill), Senate Committee on Agriculture and Forestry, 75th Cong., 1st sess. (1937), p. 2.

[10] The history of this struggle is detailed at length in the *Hearings* of the Joint Congressional Committee to Investigate the Tennessee Valley Authority (1938). The meaning of the split, in terms of its consequences for policy within the organization, has never been fully recorded. It is especially unfortunate that a study made by Professor Herman Finer of the administrative history of the Authority, undertaken in 1937 under the auspices of the Social Science Research Council, has remained unavailable.

[11] William Pincus, "Shall We Have More TVA's?," *Public Admin. Rev.*, V (Spring, 1945). *Cf.* also C. Herman Pritchett, *The Tennessee Valley Authority* (Chapel Hill: University of North Carolina Press, 1943), pp. 219 ff.

[12] Like all such statements, this seems to imply a Machiavellian motive. That is not intended; nor is it suggested that this is the sole, or indeed, any "reason" for the TVA action. The event has, however, an objective meaning and significance for the TVA's organizational strength.

[13] D. E. Lilienthal, "The TVA: A Step Toward Decentralization," an address before the University of California, Berkeley, Calif., Nov. 29, 1940 (mim.).

[14] From President Hoskins' annual statement to the Board of Trustees of the University of Tennessee, Knoxville *Journal*, August 12, 1942.

[15] See report of the Tennessee Valley governors on TVA in *Hearings* on S. 555 (Missouri Valley Authority), subcommittee of Senate Committee on Commerce, 79th Cong., 1st sess. (April, 1945), pp. 34–38.

[16] See below, pp. 145 ff.

[17] See below, chap. vii.

[18] See Avery Leiserson, *Administrative Regulation: A Study in the Representation of Interests* (Chicago: University of Chicago Press, 1942).

[19] See for example Harold J. Laski, "Discretionary Powers," *Politica*, I, nos. 1–4 (1934–1935), pp. 274–285; also Marshall E. Dimock, "The Role of Discretion in Modern Administration," *Frontiers of Public Administration* (Chicago: University of Chicago Press, 1936).

[20] Discussed in chaps. iii–vi.

[21] For further discussion of the implications of the grass-roots policy for U. S.–TVA relations, see below, pp. 74 ff.; also chaps. v and vi.

[22] Reference here is to the possible financing of development activities, such as the agricultural and regional studies program, out of power revenues, as these reach surplus proportions.

[23] Cf. Robert K. Merton, "The Role of the Intellectual in Public Bureaucracy," *Social Forces*, XXIII (May, 1945), 413.

[24] See C. I. Barnard, *Dilemmas of Leadership in the Democratic Process*, Stafford Little Lectures (Princeton: Princeton University Press, 1939), p. 11.

[25] *Ibid.*, p. 10.

[26] This is not yet clear to all participants, because the public power controversy is still important, and allegiance to TVA is defined in terms of its role in advancing public power. Support of the authority method in other areas derives much of its strength from the same sources which originally shaped the structure of TVA: the *avant-garde* quality in terms of national politics and the need for flexibility in carrying forward the electric power controversy.

[27] That this type of commitment is important is supported by the assertion of one TVA official that a leading proponent of the extension of civil service control cannot understand the TVA viewpoint on an autonomous agency civil service because of his long-standing professional commitment to the U. S. Civil Service Commission.

[28] See the two sets of *Hearings* on S. 555, "A Bill to Create a Missouri Valley Authority," Senate Committee on Commerce and Senate Committee on Irrigation and Reclamation, 79th Cong., 1st sess. (1945).

[29] *Hearings* on S. 555, Senate Committee on Commerce (April 18, 1945), pp. 122 ff.

[30] Including a judicious measure of lobbying, by no means, of course, unique to TVA among governmental agencies.

[31] *Hearings* on S. 555, Senate Committee on Commerce (April 23, 1945), p. 344.

Part Two

AN ADMINISTRATIVE CONSTITUENCY

CHAPTER III

TVA AND THE FARM LEADERSHIP: THE CONSTRUCTION OF AN ADMINISTRATIVE CONSTITUENCY

In adhering to the memorandum of understanding, the Authority is not protecting simply an institution but also the principle of our concept of democracy. If there had been no land-grant colleges in existence in 1933, in order to carry out an integrated program in the Tennessee Valley region the Authority would have found it necessary to recommend and support the establishment of such institutions.

DIRECTOR OF AGRICULTURAL RELATIONS,
TENNESSEE VALLEY AUTHORITY (1941)

THE PRECEDING chapter has been devoted to an interpretation of the grass-roots idea as official TVA doctrine. Such an analysis may stand by itself in some respects. But if we ask for the meaning of a policy in terms of its consequences in action, we must trace the history of the events within which the policy is presumed to have been effective. Moreover, it is impossible otherwise to document the hypothesis suggested above that the grass-roots policy functioned for TVA as a mechanism by which the agency achieved an adjustment to institutional forces within its area of operation.

Among the varied efforts of the TVA to execute its responsibilities under the guiding principle of grass-roots administration, the Authority's agricultural program is probably the outstanding example. This activity is a major phase of the TVA's work, and accounts for most of the funds expended under contractual arrangements with local agencies.[1] The agricultural officials within the Authority constitute the most vigorous proponents of the grass-roots approach, and it is in this field that the possible implications of the administrative principle have been most clearly worked out in action.[2] We therefore turn to this major phase of TVA's program as a case study of the policy of executing a regional program with and through existing local institutions.

TVA'S AGRICULTURAL RESPONSIBILITIES

The TVA agricultural program derives in theory from the idea of a unified development of the watershed of the Tennessee River, whereby water control on the land complements the control of water in the river {86} channel. With an average rainfall in the Valley of 52 inches, rising

81

in some areas to 85 inches, a long history of erosion, and a farm economy which has emphasized the production of clean-tilled crops such as corn and cotton, the TVA links its agricultural activities to over-all responsibilities for the protection of the watershed through the storage of water in the land.

In practical terms, however, it was the existence of the nitrate plants at Muscle Shoals, Alabama, which determined the primary content of the Authority's agricultural program. These plants were erected during the First World War and—in line with provisions of the National Defense Act of 1916—various efforts were later made, without success, to put them to peacetime use for the benefit of agriculture. Though the potentialities of the installations for commercial operation were subject to prolonged debate, pressure for the utilization of Muscle Shoals as a fertilizer project was persistent. In 1921, the significance of Muscle Shoals was formulated by the American Farm Bureau Federation:

> Nitrate Plant No. 1, Nitrate Plant No. 2, and the Wilson Dam together constitute what *is probably the greatest single conservation activity of our government.* This entire project should be viewed in the same light as an irrigation project, a forest reclamation activity, or a levee or drainage problem. *Its great purpose in peace times, was to assist in maintaining our soil fertility and consequently in the adequate protection of food for our increasing millions.*[3]

The reports of President Coolidge's Muscle Shoals Inquiry (1925) and President Hoover's Muscle Shoals Commission (1931) lent support to the drive for quantity production of fertilizers at the Shoals, although there was little response from private industry for operation of the plants. Efforts made to lease the properties on the basis of commitments for large-scale production of nitrogenous fertilizers were unsuccessful. Moreover, the chief supporter of government operation, Senator George W. Norris, had little confidence in fertilizer potentialities of the project.[4] {87} It soon appeared that the real prize at the Shoals, both for those who advocated government operation and for the private companies whose bids were considered, was the electric power generated at Wilson Dam. Nevertheless, the fertilizer aspect played an important role in the controversy up to the last. Bids of the Ford Motor Co. and the American Cyanamide Co. both involved the large-scale production of mixed fertilizers. These bids were supported by the American Farm Bureau Federation,[5] but much of the argument against them was based on the theory that the companies were in reality primarily interested in the available power and would not give adequate attention to the fertilizer program. The point is that the Muscle Shoals problem revolved around fertilizer as a major focal point.

Pressure for large-scale production of fertilizers continued up to within a year of the drafting of the TVA Bill. Although the earlier emphasis of the Farm Bureau on the need for nitrates was modified as prices for these products fell in the postwar period, its public position on the Muscle Shoals issue was maintained. Thus, in 1932, Edward A. O'Neal, President of the American Farm Bureau Federation, after pointing out that the price of nitrogen had fallen, stated:

> ...but we must look to the future to protect our American farmers, who use so much nitrogen, against a price for this product which existed formerly. Nitrogen per pound is still the most costly of the three main plant-food requirements.... I have not surrendered the thought which was advocated ten years ago relative to the importance of Muscle Shoals along nitrogen fixation lines. I merely want to convey to the Military Affairs Committee the additional thought that this phosphoric-acid proposition, particularly the prospect of developing phosphoric acid by electrolytic processes is an additional possibility at Muscle Shoals to that which was advocated formerly.[6]

At the same hearings, O'Neal emphasized: "We didn't want any experimentation in fertilizer. We want fertilizer as a dominant factor." And on the same day a Southern Congressman called for the production at the Shoals of enough fertilizer to control the national price.

In the light of these definite statements made as late as 1932, and during the entire period of the controversy after 1920, the attitude of the Farm Bureau toward the TVA is not altogether understandable. After the election of President Roosevelt, this powerful lobby swung to the support of the Norris Bill, resulting in a double compromise of its {88} earlier position: it no longer opposed a public corporation at the Shoals and now accepted a statute which gave no guarantees as to quantity production of fertilizers.[7] It is possible that an over-all agreement with the new administration in Washington included a commitment by the Farm Bureau to support the resolution of the Muscle Shoals controversy on the terms of President Roosevelt. He was interested primarily in the public power aspect of the development and in the possibilities of regional planning, but apparently not in waging a price war against the fertilizer manufacturers.[8] At the hearings on the TVA Bill, the Farm Bureau supported the New Deal measure. Significantly, the Farm Bureau did not request a definite commitment from the new agency for a minimum production of fertilizer, either nitrates or phosphates, although its representative, O'Neal, was pressed to do so by a member of the committee.[9]

It is this history of controversy over the fertilizer issue, plus the prior existence of manufacturing installations at the Shoals, which has centered the TVA agricultural program upon the production and distribution of fertilizers. However, the TVA activity in this field has been defined as experimental and has in no sense fulfilled the initial demand for large-scale production.

Although fertilizer is the central axis of the agricultural program, that program has a more general relation to the objectives and problems of the Authority. The need to hold the water on the land through improved farming practices as a flood control measure is the basis of the Authority's interest in soil conservation. The impact of the vast dam-building program on farm lands and farm families necessitated agricultural planning for readjustment, a problem heightened by the predominantly rural character of the area. In addition, unified resource development and conservation were assigned to the Authority as correlative responsibilities. It is important to make clear that while the {89} conception of TVA, held popularly and by President Roosevelt and Senator Norris, was broad, the statutory authority provided for fairly specific activities. The Act did not grant to the TVA any exclusive or unlimited governmental powers over agricultural development in the area.

In fact the legal basis which sustains the TVA agricultural program is contained primarily in Section 5 of the organic Act of 1933,[10] which provides essentially for the production, development, and introduction of fertilizer materials. The Board of Directors was authorized,

> (a) to contract with commercial producers for the production of such fertilizers or fertilizer materials as may be needed in the Government's program of development and introduction in excess of that produced by Government plants.

This authority has not been utilized; rather, the second provision,

> (b) to arrange with farmers and farm organizations for large-scale practical use of the new forms of fertilizers under conditions permitting an accurate measure of the economic return they produce

has represented the core of the program. To implement this responsibility, permissive power was granted

> (c) to coöperate with National, State, district, or county experimental stations or demonstration farms, with farmers, landowners, and associations of farmers or landowners, for the use of new forms of fertilizers or fertilizer practices during the initial or experimental period of their introduction, and for promoting the prevention of soil erosion by the use of fertilizers and otherwise,

an authorization which linked the TVA's program to existing agricultural institutions as well as to the correlative objective of soil erosion control. Further,

> (d) the board in order to improve and cheapen the production of fertilizer is authorized to manufacture and sell fixed nitrogen, fertilizer, and fertilizer ingredients at Muscle Shoals by the employment of existing facilities, by modernizing existing plants, or by any other process or processes that in its judgment shall appear wise and profitable to the fixation of atmospheric nitrogen or the cheapening of the production of fertilizer.

> (e) Under the authority of this Act the board may make donations or sales of the product of the plant or plants operated by it to be fairly

and equitably distributed through the agency of county demonstration agents, agricultural colleges, or otherwise as the board may direct, for experimentation, education, and introduction of the use of such products in coöperation with practical farmers so as to obtain information as to the value, effect, and best methods of their use.

These provisions embrace the major framework of the Authority's work in relation to farmers and agricultural agencies. In addition, provision {90} was made for the maintenance of nitrate plant no. 2 in stand-by condition in the event that nitrogenous fertilizers were not produced, and the Board was authorized

(h) to establish, maintain, and operate laboratories and experimental plants, and to undertake experiments for the purpose of enabling the Corporation to furnish nitrogen products for military purposes, and nitrogen and other fertilizer products for agricultural purposes in the most economical manner and at the highest standard of efficiency.

Thus, while permitting considerable discretion, the Act itself specified that the production and distribution of fertilizers should be the basic contribution of the Authority to agricultural development. Other provisions of the Act relating to agriculture were, however, included. Thus, Section 4 (1) stated that the Corporation

shall have the power to advise and coöperate in the readjustment of the population displaced by the construction of dams, the acquisition of reservoir areas, the protection of watersheds, the acquisition of rights-of-way, and other necessary acquisitions of land, in order to effectuate the purposes of the Act; and may coöperate with Federal, State, and local agencies to that end.

While not exclusively an agricultural problem, the acquisition of farm properties and the need to aid in the relocation and readjustment of farm families provided further concrete activities for the agricultural organization within TVA.

Beyond these specific activities, Section 22 of the Act authorized, through the President, surveys and general plans, as well as "studies, experiments, or demonstrations" in order "to further aid the proper use, conservation, and development of the natural resources of the Tennessee River drainage basin . . . for the general purpose of fostering an orderly and proper physical, economic and social development...." And Section 23 provided that

the President shall, from time to time, as the work provided for in the preceding section progresses, recommend to Congress such legislation as he deems proper to carry out the general purposes stated in said section, and for the special purpose of bringing about in said Tennessee drainage basin and adjoining territory in conformity with said general purposes (1) the maximum amount of flood control; (2) the maximum development of said Tennessee River for navigation purposes; (3) the maximum generation of electric power consistent with flood control and navigation; (4) the proper use of marginal lands; (5) the proper method of reforestation of all lands in said drainage basin suitable for reforestation; and (6) the economic and social well-being of the people living in said river basin.

While these general welfare sections of the TVA Act serve to enlarge the conception of the agency's objectives, they do not in fact grant broad {91} powers of action. This has been recognized by the Authority's leadership, resulting in a strong tendency to avoid basing any major activities upon the authority contained in Sections 22 and 23. The agricultural program in particular is rooted in the fertilizer provisions of Section 5. This does not mean, however, that a broad conception of the economic and social well-being of the rural population has not directed the administration of the program. On the contrary, the TVA exercises its discretion in the spirit of Sections 22 and 23, as these are variously interpreted.

APPROACHING THE GRASS ROOTS

The first appointee to the TVA Board was its chairman, Arthur E. Morgan, an engineer of wide reputation and broad social vision. As engineer and educator, he combined a practical knowledge of water control problems with an active interest in the general problems of social reconstruction. His Antioch plan in college education was widely known, and he had developed a philosophy of responsible public service which was to play a major role in laying the foundations for the Authority's personnel system. Morgan's selection by President Roosevelt preceded the enactment of the TVA statute, and he participated in drafting some of its provisions.

As chairman and first appointee (though the chairmanship was later revealed to be of little significance), Morgan was consulted by the President on the appointment of his codirectors. The President required that one of the directors be a Southern agriculturist. Morgan approached representatives of the Extension Service of the Department of Agriculture for suggestions as to a likely candidate from the Tennessee Valley region. As a result, the President of the University of Tennessee, Dr. Harcourt A. Morgan, was nominated by the President for the six-year term. This was a crucial selection, for it materially determined the nature of the discretion to be exercised by the new agency in its agricultural program, with attendant consequences for other phases of the agency's structure and activities. Upon reflection, a few years later, A. E. Morgan rebuked his own naïveté in having felt that elements in the Department of Agriculture consulted by him would consider professional competence rather than tangential organizational and political motives in suggesting a candidate for the TVA Board. However that may be, it seems clear that the institutional background of H. A. Morgan, particularly as a leader in the land-grant college extension service system, decisively affected the course of the TVA agricultural program and the special role of the Agricultural Relations Department among the administrative {92} units of the Authority. This is the opinion of informed members of the TVA staff.

The influence of H. A. Morgan upon the agricultural program was assured by the early division of functions among the TVA directors.[11] This followed an initial disaffection within the Board which moved Directors D. B. Lilienthal and H. A. Morgan to attempt to restrict the activities of Chairman A. E. Morgan. By this plan, adopted by the Lilienthal–H. A. Morgan majority in August, 1933, A. E. Morgan was given responsibility primarily for the general engineering program; Lilienthal was to be in charge of all matters pertaining to the Authority's power development; and H. A. Morgan was assigned supervision of all matters relating to agriculture, agricultural industries, and rural life, as well as of the chemical engineering program. Thus the Authority's power to exercise discretion in relation to the broad fertilizer and soil conservation program was concentrated in the hands of the former president of the University of Tennessee.

Dr. Morgan was the only representative of the Tennessee Valley region on the TVA Board of Directors. It is fair to say that he represented the relatively conservative institutional forces in the area, as those were expressed in the established agencies of state and local government. Dr. Morgan brought to the TVA a long experience with agricultural work in the South, together with corresponding professional commitments to those agencies with which he had been so long associated. It is he who is the recognized source within the Authority of the grass-roots approach as a moral enterprise, although public formulations have come primarily from Lilienthal. As a representative of the area, moreover, Dr. Morgan brought with him the possibility of mobilizing considerable support for the TVA from established institutions in the region. It is freely stated among certain circles within the TVA that the H. A. Morgan–Lilienthal bloc was a "log-rolling" enterprise, whereby Lilienthal received support for the electric power program in exchange for his support of the fertilizer program. Whether this was so or not, in terms of conscious behavior, is certainly open to question, but it is not unreasonable as an interpretation of the objective meaning of the bloc. Dr. Morgan's connection with the agricultural extension services, organizations ramified through every county in the watershed, represented a formidable factor which might conceivably turn the scale of popular opinion from support or indifference into antagonism. An extension service official in one of {93} the Valley states stated privately in 1943 that "the TVA investigation of 1938 would have turned out differently if the extension service had been alienated. Some witnesses would have testified differently." Another extension service official said, in 1942, "if the TVA ever tried to go out on its own, by-passing the extension service, the next appropriation bill would not come out of committee."

It would not be surprising, therefore, if Lilienthal, recognizing that the national political implications of TVA were linked above all to its electric power program,[12] and that this program was facing enormous practical and constitutional difficulties, might have been willing to agree to the delegation of discretion in the agricultural program to Dr. H. A. Morgan. This alignment eventually spelled the doom of such development programs, supported by A. E. Morgan, as were not acceptable to the land-grant college group. These included an emphasis on self-help coöperatives, subsistence homesteads, rural zoning, and broad regional planning.

The TVA was moving into the agricultural field at a time of transition in which earlier tendencies of the Department of Agriculture to extend its activities from education to action were being increased and made explicit by the new ventures of the New Deal administration. Established agencies, notably the land-grant colleges, took heed of possible threats to their institutional prerogatives, especially those which might block their nearly exclusive avenue to the farm population through the county agent organization. This concern was expressed through the powerful Association of Land-Grant Colleges and Universities, organized in 1887 for consideration of the problems and interests of the officials within the state experiment station and extension service network. In addition, the closely related American Farm Bureau Federation supplemented the activities of the Association from its vantage point as a mass-pressure group. In 1933, while plans for the initiation of the TVA agricultural program were being developed, the 47th Convention of the Association was held, at which "there was much discussion of the work of the Authority and the relations of the Land-Grant Colleges to it. A committee was appointed to confer with the Authority 'in regard to such features of its work as naturally fall within the field of work of {94} institution members of this Association.'"[13] With H. A. Morgan in charge of the TVA's agricultural program, the committee could find little difficulty in arriving at a satisfactory arrangement.

This was not so with respect to other new agencies, however. In 1934, at the 48th Annual Convention, a report of a "Special Committee on Duplication of Land-Grant College work by New Federal Agencies" was heard. This committee pointed to "ample evidence of duplication (which) invades long-established fields of the land-grant colleges in extension, in research and possibly in resident teaching."[14] Examples cited were the Federal Emergency Relief Administration's rural home demonstration work and the research, demonstration, and extension work of the Federal Soil Erosion Service. It is significant that preliminary negotiations between the latter agency, then in the Department of the Interior, and Chairman A. E. Morgan of the TVA, had been initiated, looking toward a joint program; but the proposal was rejected by H. A. Morgan, setting the pattern for a selective approach to the land-grant college system as the primary avenue of coöperation. This action of H. A. Morgan followed closely the recommendations of the Special Committee on Duplications that "this Association reaffirm its willingness to make available the trained and experienced personnel, the extensive organization and the benefits of the long experience of the land-grant colleges as the agencies through which Federal-State coöperation should function in scientific and educational service to industry, agriculture, and rural life."[15]

After two original conferences of representatives of the Valley states colleges and the TVA in 1933, and a third in July, 1934, a definite pattern emerged within which future relationships would be ordered. The seven Valley states colleges[16] selected representatives to participate in joint planning activities, and a correlating committee was created, consisting of three officials, one a joint representative of the Valley states colleges, one from TVA, and one from the U. S. Department of Agriculture. The original committee consisted of Dean Thomas P. Cooper {95} of Kentucky, Director J. C. McAmis of the Department of Agricultural Relations, TVA, and Dr. C. W. Warburton, director of Extension Work in the U. S. Department of Agriculture. The selection of Dr. Warburton insured amicable relations, since the entire correlating committee represented the extension service outlook. Later, another representative of the Department was to reflect underlying changes in organizational emphasis.[17]

THE MEMORANDUM OF UNDERSTANDING

The correlating committee succeeded in bringing the agricultural activities of the TVA into line with the established relationships between the Department of Agriculture and the land-grant colleges. In effect, this meant the establishment of a major part of the TVA agricultural program upon a grant-in-aid foundation whereby the TVA supplied funds for the execution of programs by the states, following the pattern used by the Department in administering the Morrill, Hatch, and Smith-Lever Acts. The basic document establishing the three-way relationship was completed late in 1934 as a memorandum of understanding signed by the Chairman of TVA, the Secretary of Agriculture, and the presidents of the Valley states agricultural colleges. This agreement was framed by the correlating committee to provide a "systematic procedure for a coördinated program of agricultural research, extension, and land-use planning within the region of the Tennessee Valley Authority." The correlating committee was given formal status, and the employment of an executive secretary to be jointly financed, was authorized. Dr. Carleton R. Ball was chosen for this position, with headquarters in Washington. A system of "state contact officers" was also provided for, but never implemented.

The core of the agreement was contained in the provision that proposals for joint coördinated activity on the part of any of the participants would be channeled through the correlating committee. In this way, the regional implications of proposed projects would be given due consideration. Beyond this formal objective of the memorandum of understanding, however, must be considered its implications for the organizational position of those involved. These implications have been formulated by informed participants along the following lines:

1. The three-way agreement was a device for the control of TVA—or at least its relevant activities—by the Department of Agriculture, which was at that time dominated by men committed to the land-grant college {96} system. There was some apprehension that the TVA might embark on a program of its own, or with some other agency, a fear which found some support in the attempt of A. E. Morgan to come to an agreement with Secretary Ickes of the Department of the Interior. While there was little to be feared from H. A. Morgan and the organization he built inside the TVA, a formal agreement would bolster the H. A. Morgan group if pressure should arise from other quarters in TVA for a new approach. The formal agreement established a precedent and a commitment, and a set of continuing relationships; these would offer a considerable advantage in any controversy which might arise.

2. With the TVA securely within the land-grant college framework, the agreement could serve as a bulwark against such new agencies as might wish to work in the Valley outside the framework set by the colleges. While little recourse would then be possible, still a commitment was obtained from the Department of Agriculture as a whole, so that the correlating committee procedure could be used as a point of entry and debate, a formal basis from which objections to the new action agencies—as opposed to agencies primarily engaged in research—might be lodged. One TVA staff member said that "the extension service was scared to death of the trend going toward the action agencies, and they wanted to stop it in the Valley."

It is noteworthy in connection with this interpretation that in reality the state organizations, that is, the land-grant colleges, were the strongest elements among the participants in the agreement. The TVA representatives, as will be elaborated below, actually had the interests of the colleges at heart. Having prior commitments to the land-grant college system as such, the TVA was not an independent organization bargaining freely in the light of its own interests and program. The position of the Department of Agriculture was similarly conditioned since its representation was effected through its Office of Coöperative Extension, which had been unable to retain a status independent of the states. The Office of Coöperative Extension has been a small organization, relatively helpless in respect to supervision of the states, and inescapably committed to the land-grant institutions. It is one of those organizations of which it has been said that "control runs from the state to the federal agency rather than the other way."[18] This situation is in part due to, and emphasized by, the role of the Association of Land-Grant Colleges and Universities, which tends to "constitute a more coherent body than {97} does the Department itself," and has functioned to preserve "the jurisdictional sphere of the state stations against encroachments by the research bureaus of the Departments."[19] With such considerations in mind, it seems fair to say that the memorandum of understanding cannot be comprehended as an administrative document apart from the context of informal and tangential goals and relationships within which it was written. This situation, in which the specific weight of the state organizations turned out to be disproportionately heavy, lends support to the view that these local institutions found informal representation within the TVA administrative structure through the H. A. Morgan group.

The correlating committee has gained some notice in terms of its formal objective of achieving an integrated regional agricultural program.[20] However, there is some doubt that the committee as such has functioned as a controlling device. The memorandum of understanding is implemented by contracts between the TVA and individual state universities, including a master contract and specific project agreements. These are not channeled through the correlating committee, and are not formally coördinated through any other region-wide institution. "In general," said one official intimately connected with the work of the correlating committee from its inception, "the 3-man committee has not functioned beyond calling the deans' and directors' conferences. They did have a meeting about the time the Soil Conservation Service situation got bad, but this only served to aggravate matters." A parallel interpretation was given by a member of the committee. It appears that problems of jurisdiction and liaison have constituted its primary function, although Valley-wide conferences of deans of colleges and directors of the state experiment stations and extension services with representatives of the TVA and the Department of Agriculture have been held semiannually. Minutes of the conferences indicate that many substantive and administrative problems of mutual concern have been discussed, and this has doubtless contributed to a heightened regional self-consciousness as well as to a regional exchange of information and ideas. An examination of the record of these conferences indicates that the TVA representatives have not exercised the strong leadership which their special regional responsibilities might seem to imply. The TVA representatives have {98} consciously held back, on the assumption that a Valley program must come from the states, on their own initiative. One TVA staff member (not, however, connected with the agricultural unit) contended in 1943 that "a truly regional agricultural program has not been achieved. Rather, the methods have followed the general pattern of federal-state relations in the administration of federal grants-in-aid. Such a pattern would allow for duplication among the activities of the states." In the absence of notable accomplishments in respect to the formal objectives of the three-way agreement, it would seem that the informal implications, noted above, gain increased significance.

THE DECISION FOR PHOSPHATES

As we have seen, the first major act of discretion exercised by the Authority in relation to its agricultural responsibilities was the establishment of coöperative relationships with the land-grant colleges in the Valley. Concomitantly, it was necessary to decide whether the fertilizer program of the Authority would revolve around the production of nitrates, as appeared to be implied in the Act, but was clearly not mandatory, or some other fertilizer product. On this, the position of the land-grant colleges and the American Farm Bureau Federation had already been made clear: the combination of lime, phosphate, and legumes would break the cycle of soil exhaustion brought on by the use of commercial nitrates under soil-depleting crops. These crops (including cotton, corn, and tobacco) are clean tilled, so that emphasis upon them in the agricultural economy adds to the problem of soil erosion, which in turn means increased flood hazards. It was not surprising, therefore, that, when the TVA approached the land-grant colleges for advice, the result was an exclusive emphasis on the production of phosphatic fertilizers. The contribution of TVA was to be the development and production of concentrated phosphates in order to reduce the burden of shipping costs, together with the distribution of large quantities of the new materials for practical testing and demonstration of their usefulness. The latter phase constitutes the agricultural as distinct from the chemical engineering program of the Authority.

The emphasis on phosphates has received wide support in agricultural circles, and surprisingly little criticism, in view of the striking abandonment of work on nitrates.[21] Criticism, however, was voiced to the Joint {99} Committee Investigating the TVA by the Board of Trade in Sheffield, Alabama, that the immediate needs of poor farmers can be met only by cheap nitrogenous fertilizers, since such farmers are unable to afford a program of crop rotation and nitrogen fixing through the use of legumes. Such a view finds support in the statements of some Farm Security Administration supervisors that it is impossible, in view of the pressing needs of their clients for a cash crop, to avoid making loans for the purchase of commercial nitrates. The possibility has also been suggested that the phosphate program may have been a device for turning aside government competition from nitrates, as the chief interest of the fertilizer industry. No evidence could be found to support this charge. Moreover, the fertilizer industry is less than enthusiastic about even the present TVA program, and there is evidence that pressure was brought to bear upon the Agricultural Adjustment Administration, in 1939 and 1940, to suspend the distribution of phosphates as grants-in-aid to farmers. Whatever the reasons, soil conservation or otherwise, the turn to phosphates was made, in line with the program of the land-grant colleges.[22]

EXTENT OF THE PROGRAM

In the light of the long Muscle Shoals controversy, which involved persistent agitation for the production of cheap fertilizers, it should be pointed out that the TVA has not significantly affected the over-all output of the fertilizer industry. The Authority restricted itself to an experimental program, although the initial interest in the fertilizer possibilities of Muscle Shoals centered around large-scale production of nitrates. Indeed it was the fear that the offers of Henry Ford and the American Cyanamide Co. did not include a bona fide interest in the quantity production of nitrates which was the source of considerable {100} opposition to them. It was believed that the major interest of these corporations was in the electric power possibilities of the installations. Ironically, concentration on the electric power issue appears to have resulted in limited fertilizer production at the Shoals even under public ownership. Whatever the merits of the issue, it seems somewhat anti-climactic that, after the long debate as recorded in volumenous Congressional hearings, the issue of cheap fertilizer should have been resolved by an act of administrative discretion. It is doubtful that it could have been so resolved without the consent of powerful political interests involved in the controversy. The curious elimination of the strong pressure for a large nitrate program at the Shoals may be explained in part, however, by the fact that electric power became the crucial element in the struggle over TVA as a whole, so that little attention was given to the fertilizer program by the managers of the liberal-progressive forces.

TABLE 1
DISTRIBUTION OF TVA FERTILIZERS TO JANUARY 1, 1943

	Tons
For Experiment Stations	
Ten different materials used in preliminary investigations	5,570
For Test-Demonstration Farms	
Triple superphosphate	149,810
Calcium metaphosphate	33,908
Fused rock phosphate	1,218
Total	184,936
Calcium silicate slag, a liming-material by-product	3,008
For Agricultural Adjustment Administration	
Triple superphosphate	240,198
Calcium silicate slag	341,201

SOURCE, Table 1: *Test-Demonstration Information Book*, published by the Land-Grant Colleges and Universities of the Tennessee Valley States and the Tennessee Valley Authority (March, 1943), p. 109.

Table 1 summarizes the amount of materials distributed by the Authority to January 1, 1943. Later, in connection with the resumption of nitrogen manufacture for war purposes, some nitrogenous fertilizer was produced and distributed. However, during the first decade of its operation, the TVA fertilizer program was almost exclusively devoted to phosphates. The 1939 Census of Manufactures[23] reported that the {101} fertilizer industry had produced that year 5,088,468 tons of mixed fertilizer and 4,152,642 tons of superphosphate. TVA production, while doubtless of real significance over the long run, through the introduction of new concentrated forms, has affected neither this total output nor the general price of fertilizer.

TABLE 2

SUMMARY OF PAYMENTS UNDER COÖPERATIVE CONTRACTS WITH STATE AND
LOCAL INSTITUTIONS, FISCAL YEAR, 1942

Department	Amount
Agricultural relations	$ 746,048
Commerce	78,555
Reservoir-property management	131,485
Chief engineer	612
Health and safety	71,580
Personnel	2,200
Regional studies	5,700
	$1,036,180

SOURCE, Table 2: TVA Budget Office.

TABLE 3

REIMBURSEMENTS MADE TO LAND-GRANT COLLEGES UNDER COÖPERATIVE
AGREEMENTS FOR PERIOD JULY 1, 1935, TO JANUARY 1, 1943

Institution	Total reimbursed	Master agric. contract
Alabama Polytechnic Institute	$ 699,048.76	$ 504,046.87
Georgia Universities	402,137.16	312,226.96
University of Kentucky	231,255.99	199,776.93
Mississippi State College	231,675.52	178,289.32
North Carolina State College	434,498.59	402,041.41
University of Tennessee	2,283,116.92	1,874,146.62
Virginia Polytechnic Institute	322,849.07	317,931.15
	$4,604,582.01	$3,787,459.26

SOURCE, Table 3: TVA Auditing Section.

The predominance of the agricultural program among those exemplifying the grass-roots approach within TVA is indicated in table 2, which lists reimbursements by TVA under coöperative contracts with state and local institutions in 1942. In table 3, total reimbursements to the colleges and those made under the master agricultural contracts are reported, covering 1935 to 1943. These are useful because the {102} scope of the TVA agricultural program cannot be gauged simply from a study of the bulk of materials distributed. In its educational effort, the employment of personnel has been of considerable importance, constituting the major factor in reimbursements by TVA to the colleges.

A good indication of the extent of the TVA agricultural program, as it operates through the state institutions, may be derived from the number of officials of the land-grant colleges assigned to the TVA program, with salaries ultimately paid by the Authority through a reimbursement procedure. These figures are reported in table 4.

TABLE 4

PERSONNEL ON COÖPERATIVE TVA–LAND-GRANT COLLEGE AGRICULTURAL
PROGRAM, SALARIES REIMBURSED BY TVA IN 1942

Fertilizer Operations—Test Demonstrations
  Contractual: members of college staffs
    Inside Valley states 171
    Outside Valley states 30
  Direct: Members of TVA staff 11
Fertilizer Operations—Distribution
  Direct: members of TVA staff 6
Fertilizer Operations—Program Exposition
  Direct: members of TVA staff 5
Preliminary Investigations (Experiment Stations)
  Contractual: members of college staffs
    Inside Valley states 35
Soil Inventories and Land-Use Studies
  Contractual: members of college staffs
    Inside Valley states 34
  Contractual: Bureau of Plant Industry 3
  Direct: members of TVA staff 7
Readjustment Program
  Contractual: members of college staffs
    At reservoir-affected areas 63
  Direct: members of TVA staff 5

SOURCE, Table 4: Budget of Agricultural Relations Department, TVA. Figures on TVA include both supervisory and clerical personnel.

Personnel at the professional level in the Agricultural Relations Department of TVA is limited to about thirty-one individuals. This parallels the pattern of a small headquarters staff evidenced in the relation between the Washington Office of Experiment Stations and Extension Service and the land-grant colleges. The program as a whole, however, is carried on by a network of TVA subsidized personnel stationed in every county within the watershed. Most of this personnel is {103} concentrated in the test-demonstration and readjustment programs, represented primarily by assistant county agents working out of the local county agents' offices.

TABLE 5

TVA MATERIALS INVESTIGATED BY EXPERIMENT STATIONS IN THE
TENNESSEE VALLEY TO 1943

	Tons
Triple superphosphate	3,166
Calcium metaphosphate	1,760
Calcium silicate slag	69
Fused phosphate	460
Dicalcium phosphate	32
Phosphoric acid (liquid)	23
Raw rock phosphate	15
Monocalcium phosphate	13
Potassium metaphosphate	32
Potassium-calcium metaphosphate	3

SOURCE, Table 5: *Test-Demonstration Information Book*, P. 85.

TABLE 6

TEST DEMONSTRATION FARMS, AS OF JANUARY 1, 1943

In Valley states	Unit farms	Area farms	Total
Inside Valley	6,520	12,974	19,494
Outside Valley	3,194	......	3,194
In non-Valley states	4,545	110	4,655
	14,545	13,084	27,343

Total area of 27,343 test-demonstration farms active January 1, 1943: 4,005,854 acres.

Total area of 41,951 test-demonstration farms established to January 1, 1943: 6,232,207 acres.

SOURCE, Table 6: *Test-Demonstration Information Book*, P. 107.

From the funds and personnel thus expended has been elaborated a program of plant food testing, farm management education, and population readjustment, to name only the major phases of the work.[24] The experiment stations of the land-grant colleges have made numerous tests of concentrated phosphatic fertilizers manufactured by the Authority. In 1943, it was reported that a total of over 5,500 tons of experimental {104} fertilizer materials had been subjected to such investigation. This figure is broken down in table 5. An indication of the extent to which the program reaches the farm population is presented in table 6, from which it will be noted that 27,343 farms in twenty-eight states were involved in the fertilizer distribution activities. Table 7 reports the number of families removed from reservoir areas, as an index of the extent of the readjustment activities.

TABLE 7

NUMBER OF FAMILIES REMOVED FROM TVA RESERVOIR AREAS TO MARCH 1, 1943

Area (by dam)	Total	Property owners	Tenants
Norris	2,899	1,708	1,191
Wheeler	842	61	781
Pickwick	506	126	380
Guntersville	1,182	152	1,030
Chicamauga	903	263	640
Hiwassee	261	76	185
Watts Bar	832	196	636
Cherokee	875	376	499
Chatuge	278	166	112
Nottely	91	26	65
Appalachia	22	4	18
Ocoee No. 3	0[a]		
Douglas	525	201	324
Fort Loudoun	273	91	182
Kentucky	1,327[b]	511	816
Fontana	482[c]	14	468
South Holston[d]	71	37	34
Watauga	43	28	15
	11,412	4,036	7,376

SOURCE, Table 7: Reservoir Property Management Department, TVA.

[a] Ocoee No. 3 Reservoir presented no family removal problem. The Authority acquired the land from Tennessee Electric Power Co.; in August, 1940, it was transferred to U. S. Forest Service, TVA retaining the right to flood it for Reservoir purposes.

[b] Kentucky Reservoir 51 per cent evacuated.

[c] Fontana Reservoir 40.1 per cent evacuated.

[d] Cessation of construction ordered by War Production Board.

THE TVA MACHINERY

The major administrative arm of the Authority in the field of agriculture is its Agricultural Relations Department. It is this organization which carries on the coöperative fertilizer program, particularly those aspects which directly affect the agricultural population. As we have seen (table 2), its contractual relations with the land-grant colleges account {105} for the major portion of funds used within the framework of the grass-roots approach.

Before discussing the character of the Agricultural Relations Department, some attention should be paid to its place within the general, formal TVA structure. Chart 1 (p. 106) reproduces the official organizational chart of TVA, and chart 2 (p. 107) abstracts those units which are delegated responsibilities relevant to the agricultural program. Indicated upon the latter will be noted:

{106}

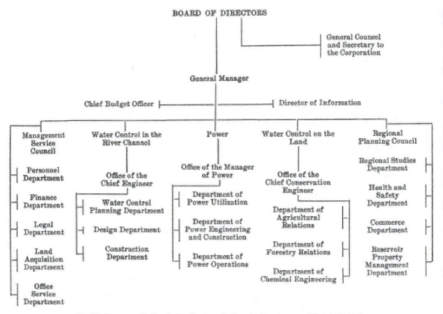

Chart 1. Tennessee Valley Authority organization chart, as approved March 16, 1942.

{107}

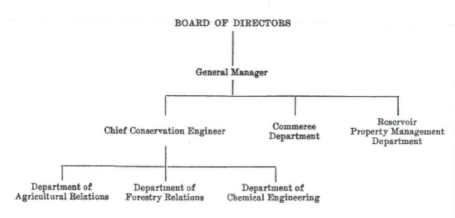

Chart 2. TVA units having agricultural responsibilities (1943).

{105, cont.}

Chief Conservation Engineer.—This post was created in 1937, concomitantly with the general reorganization of the Authority, and symbolizes the unified approach to "water control on the land." Placed in charge of three subject-matter departments, this official was assigned the responsibility of coördinating "the Authority's program of national defense, laboratory research, experimental production, and test of fertilizer and fertilizer ingredients, agricultural demonstrations and forestry conservation studies and activities, as carried on by the Chemical Engineering, Agricultural Relations, and Forestry Relations Departments..."[25] C. Niel Bass, a former official of the city of Knoxville, was appointed to this post, and has been its only incumbent. He was originally brought into the Authority as an administrative assistant to H. A. Morgan. In an appraisal of the working relationships, the position of Chief Conservation Engineer does not have the *de facto* significance which its formal status would seem to imply. Little effective control over the Agricultural Relations Department is exercised, if any was intended. On major issues, Bass tends to follow the lead of the agriculturists, as that is formulated by Director H. A. Morgan and J. C. McAmis, head of Agricultural Relations. It is believed in some quarters that Bass's appointment was in effect a means of consolidating the dominant position of the agriculturists in relation to kindred subject-matter departments. This interpretation gains strength from the history of conflict involving the early leadership of the Authority's forestry program, which was linked to the A. E. Morgan group. With the departure of E. C. M. Richards, Chief Forester, as part of the turnover of leading personnel following the ouster of A. E. Morgan, it was necessary to reintegrate the remaining members of the forestry division. The coupling of the reorganized Forestry Relations Department to the leading supporters of H. A. Morgan may reasonably be interpreted as a device chosen for this *Gleichschaltung.*[26] {108}

Department of Chemical Engineering.—This is the Authority's technical organization for research in and production of fertilizer materials. It is in direct charge of the physical plant at Muscle Shoals. A Phosphate Division conducts mining operations in the TVA phosphate properties in middle Tennessee; a Chemical Operations Division constructs and operates the commercial-size plants for fertilizer (and munitions) ingredients; a Chemical Engineering and Design Division contributes designs for the Operations Division and provides miscellaneous engineering services; and a Chemical Research and Development Division operates a laboratory-to-pilot-plant research program. This work is supplemented by coöperation with the University of Tennessee in the maintenance of a Chemical Engineering Research Laboratory and the Coöperative Chemical Research Unit at the University's Experiment Station. The basically technical character of the functions of the Chemical Engineering Department was underlined in 1941, when the Commerce Department was made responsible for coördinating the industrial research activities of the Authority and for their articulation with the research activities of other agencies.[27]

Department of Forestry Relations.—Chiefly through its Watershed Protection Division, this organization is related to the agricultural program by virtue of its responsibility to coöperate with other local, state, and federal agencies for erosion control, reforestation, farm forestry, and forest protection. Coöperation with the state extension service organizations has been instituted, though not in the same manner or with the same consequences as with the Agricultural Relations Department.[28] Technical supervision for camps operated by the Civilian Conservation Corps has been provided, and nurseries are operated to produce tree seedlings and explore and test special tree crop species. This Department also includes a Forest Resources Division, which conducts studies for developing the forest wealth of the Valley. A Biological Readjustment Division furthers the wild-life resources of the region.

Commerce Department.—This was established in its present form in 1939, by consolidation of an earlier Commerce Department, having more restricted functions, with the former Department of Agricultural Industries. Continued agricultural interest was manifested in the responsibility which the Board delegated to it to conduct research in the development of new agricultural equipment, income-producing rural electrical equipment, and the development and stimulation of {109} rural electrification education. In line with the grass-roots policy, the Department was enjoined to "carry on its activities in such a manner as to stimulate the efforts of other agencies in the region on a permanent basis, either through coöperative formal or informal understandings or agreements, and with due regard for the needs and facilities of the coöperating agencies."[29]

Reservoir Property Management Department.—Responsible for the administration of TVA-owned lands, this organization maintains relations with the state extension services, which—especially in the organization of the local Land-Use Associations, through which rents are collected and conservation practices advanced—aid in dealing with those farmers who rent Authority land for agricultural purposes.

Formal Structure of the Agricultural Relations Department.— Operating on a budget of well over one million dollars yearly, this unit within TVA is the foremost representative of the grass-roots approach, both in its own vigorous espousal of the doctrine and in view of its central role in the TVA's relation to the farm population. As we have already noted, the headquarters staff is small, and no TVA field offices are maintained, the entire operating load being carried by the land-grant colleges of the Valley area. The bulk of the department's available funds is disbursed to the colleges virtually on a grant-in-aid basis. Chart 3 (p. 110) reproduces the official organization chart of the department.

The Test-Demonstration and Reservoir Adjustment Divisions bulk largest in the agricultural program. The former is responsible for supervision of the arrangements with the land-grant colleges under which the Authority's fertilizer products are supplied to farmers for conducting unit and area test-demonstrations on representative farms. The TVA staff for this work numbers only four professional and seven clerical personnel, while a total of 201 TVA-subsidized supervisors, specialists, and assistant county agents are employed by the colleges.[30] The Reservoir Adjustment Division is responsible for providing assistance and advice "with regard to problems of land use, and to relationships involving or affecting agricultural institutions and groups where such problems are associated with land acquisition, the readjustment of population in reservoir-affected rural communities and the management of TVA-controlled lands."[31] Four supervisory officials are on the Agricultural {111} Relations Staff at Knoxville, while sixty-three are employed through the land-grant colleges to work in reservoir-affected areas.

{110}

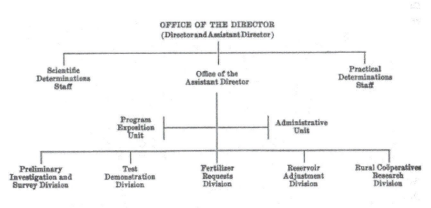

Chart 3. TVA Department of Agricultural Relations organization chart as approved December 10, 1940.

The Program Exposition Unit is a small group providing educational materials for farm groups and the general public. The Preliminary Investigation and Survey Division deals primarily with the experiment stations in relation to soil surveys and controlled tests of fertilizer products. The Fertilizer Requests Division is a service organization, facilitating the routing, dispatching, and accounting control of fertilizer material distributed in connection with the test-administration program. A small Rural Coöperatives Research Division'"[32] provides assistance in relation to those programs of the Authority which involve the development of coöperatives, in particular the county soil associations and the rural electric coöperatives. Relative importance of divisions is shown in table 8.

TABLE 8

Allotment of Funds, by Program and Method of Disbursement, within
TVA Agricultural Relations Department (1943–1944)

Program	Contractual	Direct
Test demonstration	$ 507,000	$ 29,000
Readjustment and relocation	372,000[a]	25,000[a]
Fertilizer requests	....	19,000
Program exposition	....	25,000
Soil and fertilizer investigations	90,000	28,000
Soil inventory	87,000	18,000
Coöperatives	....	25,000
General and administrative	....	79,000
Total budget: $1,304,000	$1,056,000	$248,000

SOURCE, Table 8: Proposed departmental budget, TVA, Department of Agricultural Relations, 1943–1944.

[a] Approximate.

Character and Role of the Agricultural Relations Department.—
Like Harcourt A. Morgan on the TVA Board of Directors, the Agricultural Relations Department functions as an institutional representative of the Tennessee Valley region within the Authority. In particular, it functions as a representative of a special local leadership: the land-grant colleges and the established farm leadership. An interpretation which ignores this relation cannot help being weak and unrealistic; on the other hand, to make use of this principle is to be in line with those within the TVA who are well informed. {112}

The department is a reflection of the views and personality of Director H. A. Morgan. This is assured by the close personal relation of the department head to Dr. Morgan and by the institutional origins of the other leading personnel who administer the agricultural program. Not unnaturally, the former president of the University of Tennessee chose a member of the university farm management staff, J. C. McAmis, to be his second-in-command in developing the Authority's agricultural activities. McAmis joined the TVA staff in 1933 and since then has been in continuous charge of the Department of Agricultural Relations and its earlier homologues. Like Dr. Morgan, he is a vigorous advocate of the land-grant college system, and a strong defender of the South against encroachments by those inclined to come with what appears to him to be more fervor than understanding. His loyalty, and that of others on the staff, to "Dr. H. A." is often expressed and emotionally buttressed by the evangelical halo which surrounds Dr. Morgan's attitude to the problem of "man and phosphate."

The leadership of the department is concentrated in the hands of four or five men whose professional roots lie primarily in the Tennessee extension service. Thus the major phases of the TVA fertilizer program, so far as it is brought directly to the farmer, are administered by a group of men whose attitudes and social commitments have been framed by forces within the Valley area. It is a group which is strongly conscious of the need to preserve the integrity of the extension service organizations, on the state level, in all relations with the TVA.

The institutional origins of this leading group imply a set of basic attitudes of great significance in the administration of governmental programs. These attitudes, among others, relate to the Negro population in the South and to farm tenancy. There is evidence that the typical position of the TVA agriculturist on the former is one of white superiority. This is supported by references to "good and bad niggers," antagonism toward those who "meddle" in such issues, as well as by the typical refusal to accord to Negroes such symbols of courtesy and mutual respect as the title "Mr." On the problem of land tenancy, the prevailing attitude is one of paternalism, with the assumption that landlords should "take care of" their tenants, who are generally deemed to be satisfied with their lot. The regional pattern of relative unconcern for the special position of the Negro population is carried out in the exercise of discretion in utilizing local institutional resources. The Negro agricultural colleges have had no place in the TVA fertilizer program, and the opportunity to increase the number of Negroes on the staff of the extension {113} program in the field has not been utilized.[33] Negro farmers have been involved in the program through the regular county-agent system working out of the white universities, but no special effort has been made by the TVA agriculturists to strengthen the Negro colleges. Traditional attitudes toward tenancy may be related to the hostility of the department to the tenant-oriented Farm Security Administration.[34]

These responses of the TVA agriculturists are representative of those prevailing within their area of operation. Consequently, in one view, they may be considered to legitimately implement the grass-roots method. Yet it may be suggested that the Authority might have picked a leadership with a more forward-looking approach to pressing rural problems, and cautiously attempt to induce the local institutions to show greater concern for the relatively dispossessed elements. It might be suggested that the Authority need not have chosen men who would simply reflect prevailing institutional attitudes; a local leadership need not imply a leadership totally acquiescent in established inequalities and their supporting codes. Indeed, in the early days of the Authority, some effort was made to make official contact with rural leadership tending in a different direction. But this did not come on the initiative of the agriculturists within the TVA and came to nothing.

If, however, the TVA grass-roots policy, in its agricultural phase, is viewed in the light of its organizational function, the consistency and reasonableness of the character of the Agricultural Relations Department become clear. It was suggested in the preceding chapter that the meaning of the grass-roots doctrine might be interpreted significantly in terms of the need of the TVA to adjust, to make its peace, with at least a major segment of the local institutions. As the TVA is an imposed {114} agency—born in the midst of and faced with a long struggle against powerful interests in the electric power industry, cognizant of the potential sources of hostility in sectional self-consciousness and states-rights sentiment, and committed on a national scale to the predominant importance of its electric power program—the conditions of its existence easily support the interpretation that the fertilizer program was turned over to the existing leadership in the area in exchange for at least tacit support of the power program. Whatever the motives of the leading participants, this appears to be the objective import of the organizational relationships which were established.

The self-consciousness of the Agricultural Relations Department in respect to its social origins is remarkable, and may be referred to a defensive psychology generated by being part of a larger organization with quite different roots. The department chief once said, significantly, "——, myself, and—— are products of a system—you can take your coat off when you come into TVA, but you can't take your hide off." Faced with the need to defend the region and its institutions against alien elements, it is not surprising that the TVA agriculturists should have developed a special loyalty within the organization. This defense is especially important because social origin does not appear to be adequate as an explanation for the institutional role of this group inside TVA. Other farm agencies, such as the Agricultural Adjustment Administration, Farm Security Administration, and Soil Conservation Service have recruited professional personnel from the local extension service staffs, without, so far as is known, developing the same self-consciousness. This may be a matter of training in service. It is probable that H. A. Morgan and J. C. McAmis, having uncontested control over the agricultural program, so defined the situation for new personnel that the question of defending the integrity of local institutions became paramount in the minds of the basic agricultural staff. In other agencies, such as Soil Conservation Service, a determined effort is made to reorient personnel so that the viewpoint of the leadership may be suffused through the ranks. Some SCS men, recruited from extension service, reported that they "really learned about soil conservation since joining the SCS." An analogous statement, with implied deprecation of the extension service, would not find favor in the eyes of the TVA agricultural leaders.

The position of the Agricultural Relations Department is reinforced by the theory of TVA central management which values decentralization within the Authority. The General Manager's Office is committed in theory to permit subject-matter departments a wide measure of discretion {115} in the administration of their programs. In one sense, this only officially validates the strong precedents, established in the early days of the Authority, for a "hands-off" attitude toward the agricultural program. Doubtless, however, the fact that the department has virtually its own representative, H. A. Morgan, on the three-man Board of Directors must underlie all other considerations. Some elements in and around central management have nevertheless attempted to bridle the department with tentative and somewhat hesitant attempts to institute controls from above. "We tried," said one man close to central management, "putting someone above him, below him, and alongside him." The "him" refers to the Director of Agricultural Relations, and the other persons to, respectively, the Chief Conservation Engineer,[35] an assistant director, and a consultant. None of these efforts was effective, however. It is significant that in at least one case the aggressive and evangelical personality of the department chief was able to shape the viewpoint of a young administrative assistant, who is supposed to have been sent in by central management. As a consequence, when this man was later transferred to a strategic post in central management, it was possible for the department to have an ideological representative in the General Manager's office, writing memoranda on controversial matters.

On the whole, however, the position of central management is not unambiguous. It is by no means clear that the General Manager himself, keenly aware of the problems involved, is (1943) behind or supports all attempts by elements in his office to strengthen formal control over the Agricultural Relations Department. Thus, when in 1943 the TVA Budget Office sought to bring about tighter control over contractual relations with the land-grant colleges, and to strengthen the role of central management in the administration of those contracts, there was some doubt as to the seriousness of the top leadership itself. This was not lost to the Agricultural Relations Department. In commenting on the report of the Budget Office, one of the leading agriculturists suggested:

> It seems to me that the important thing for us to do first is to find out to what extent the General Manager's office considers this report of importance. How much attention should we give it? It may be that the General Manager's office knows certain things should be done about which we cannot be informed.

It appears that the General Manager and his closest associates are more inclined to deal in personal and informal terms with department heads than to rely on formal controls.

The Agricultural Relations Department, however, is not unwilling to use formal mechanisms to further its special interests. Thus, in 1943, the {116} department tried to have central management promulgate an official administrative code for the "coördination of the activities and relationships pertaining to agriculture" which in general would give to the Agricultural Relations Department the power to coördinate all regional development programs undertaken by the Authority and affecting agriculture in any way. This would include programs carried on by the TVA Commerce Department such as the processing of food and fiber products. It would channel through the department all contact by the Authority with farm people. Its significance is that the strong emphasis of the proposed code on dealing with the local institutions would assure the position of land-grant colleges. Relationships instituted voluntarily and enthusiastically by the agriculturists in TVA would become virtually an enforceable administrative rule for other TVA departments.[36]

Rumblings of a more serious nature, sufficient to shake the equanimity of the agriculturists, have occurred, however. In September, 1943, at a regular joint conference of the Board of Directors and representatives of management, Director Lilienthal is reported to have expressed concern over the economic and social phases of agricultural development in the Valley. From the point of departure of the report of a Congressional committee on industrialized farming in California, the problem of possible changes in the pattern of farm ownership in the Tennessee Valley, and especially the place of the family farm, was discussed. It was suggested that the present TVA agricultural program might be inadequate in face of such possible changes. The background of the TVA agriculturists has inhibited serious consideration of such problems, so that it is safe to say that the department was caught off guard by the remarks of the then TVA Chairman. Lilienthal, it appears, considered it a function of TVA management to provide leadership for the region. But the agriculturists inside the organization incline to the view that they "make no effort to lead development in any particular direction" and that, indeed, they have no special program distinguishable from that of the land-grant colleges in the field of agriculture. From trends observable in 1942–1943, however, it is probable that the department will find it expedient to undertake a broader measure of initiative, particularly in the development of farmers' coöperatives.[37]

---

[1] See table II, p. 101 below.

[2] It should perhaps be said that a number of officials within TVA would consider the agricultural program a somewhat extreme rendering of the grass-roots method. However, the central place afforded to it here is in fact a reflection of its equally important role within TVA itself. Even were that not so, the selection of an extreme example is justifiable on theoretical grounds. It is precisely the extreme case, in which counteracting effects have been minimized, which permits an examination of the principle laid bare.

[3] Reported in a brief of the American Farm Bureau Federation submitted to the Congressional Joint Committee to Investigate the Tennessee Valley

Authority, Nov. 21, 1938 (mim.), p. 2. Emphasis in the original. The good faith of the American Farm Bureau Federation with regard to the fertilizer problem has been called in question at various times; but this does not vitiate the symptomatic significance of the Federation's demands for fertilizer production. Moreover, the national Farm Bureau's active support of the Soil Fertility Bill (See *Hearings* on S. 1251, Senate Committee on Agriculture and Forestry, 80th Cong., 1st sess., May, 1947), against the opposition of the fertilizer companies, operates to discount such criticism.

[4] See C. Herman Pritchett, *The Tennessee Valley Authority* (Chapel Hill: University of North Carolina Press, 1943), pp. 9–13.

[5] For the record of the Farm Bureau's positions on these offers and related matters, see *Hearings*, House Committee on Military Affairs, 72d Cong., 1st sess. (January 7, 1932), pp. 294–298.

[6] *Muscle Shoals Hearings*, House Committee on Military Affairs, 72d Cong., 1st sess. (January, 1932), p. 266.

[7] Testimony before the Senate Judiciary Committee during its lobby investigation of 1929–1930 revealed (see p. 2892 of the *Hearings*) that the Farm Bureau was a consistent enemy of the attempts of Senator Norris to achieve a program of public ownership as a solution to the Muscle Shoals problem. In 1928, the Farm Bureau supported President Coolidge's veto of a resolution establishing a Muscle Shoals Corporation, calling the resolution "an abject surrender of what agriculture has been fighting for at the Shoals in that the production of fertilizers for experimental purposes will not provide agriculture with the highly concentrated complete fertilizers which are necessary" (p. 2925).

[8] The Farm Bureau has been accused of "selling out" the farmers on the Muscle Shoals issue. See, for example, *National Farm Holiday News*, organ of the left-wing Farm Holiday Association, October 30, 1936. However, it appears that the Farm Bureau abandoned its demands for quantity production of fertilizer at the Shoals only at the beginning of the Roosevelt administration.

[9] *Muscle Shoals Hearings*, House Committee on Military Affairs (April, 1933), p.85.

[10] 48 Stat. 58 (May 18, 1933).

[11] See the "Memorandum on Organization," August 3,1933, published in *Hearings*, Joint Committee to Investigate the Tennessee Valley Authority, 75th Cong., 3d sess. (August, 1938), pp. 105–107.

[12] "At the heart of the TVA controversy is its power programs. Attacks upon the Authority are sometimes made in terms of personalities, in terms of its navigation and flood-control program, and in terms of its fertilizer work or other activities, but I think the Committee will find that, at bottom, all of the attacks upon the Authority stem from 'power'." Testimony of David E.

Lilienthal, *Hearings*, Joint Committee to Investigate the Tennessee Valley Authority, 75th Cong., 3d sess. (August, 1938), p. 149.

[13] Carleton R. Ball, *A Study of the Work of the Land-Grant Colleges in the Tennessee Valley Area in Coöperation with the Tennessee Valley Authority*, prepared under auspices of the Coördinating Committee of the USDA, the TVA, and Valley States Land-Grant Colleges, 1939, p. 12. (No imprint.)

[14] *Proceedings* of the 48th Annual Convention of the Association of Land-Grant Colleges and Universities, Washington, D.C. (Nov. 19–24, 1934), p. 240.

[15] *Ibid.*, p. 241.

[16] Alabama Polytechnic Institute, Auburn, Ala.; University of Georgia, Athens, Ga.; University of Kentucky, Lexington, Ky.; Mississippi State College, State College, Miss.; North Carolina State College, Raleigh, N.C.; University of Tennessee, Knoxville, Tenn.; Virginia Polytechnic Institute, Blacksburg, Va.

[17] See below, p. 170.

[18] J. P. Harris, in foreword to V. O. Key, *The Administration of Federal Grants to States* (Chicago: Public Administration Service, 1937), p. xii.

[19] Key, *op. cit.*, pp. 179, 201. See also G. A. Works and Barton Morgan, *The Land Grant Colleges*, Staff Study No. 10, Advisory Committee on Education (Washington: U. S. Govt. Printing Office, 1939), p. 107.

[20] See John M. Gaus and Leon O. Wolcott, *Public Administration and the U. S. Department of Agriculture* (Chicago: Public Administration Service, 1940), p. 371.

[21] The TVA was directed in its Act to maintain nitrate plant no. 2 in stand-by condition. But the phosphate emphasis apparently strongly modified even the national defense implications of the nitrate program, for TVA's chemical contribution to World War II was primarily elemental phosphorus.

[22] The TVA chemical engineers believed that the manufacture of nitrates was unfeasible with the plant taken over by the Authority at Muscle Shoals. This may have been the immediately decisive factor in reaching the decision to manufacture phosphates rather than nitrates. See *Hearings*, Joint Committee to Investigate the Tennessee Valley Authority, p. 1289. However, the decision is usually defended not on those grounds but on the theory that regional needs were best served by a phosphate program. It would be difficult to understand why the Authority did not come to Congress with a program for such alterations in the fertilizer plant as would make nitrate production technically feasible (since that was widely thought to be the responsibility of the Authority) were it not that the decision was taken on nonengineering grounds. In this connection, H. A. Morgan reported that in 1933 he held "a series of conferences with officials in the

Department of Agriculture, representatives of the experiment stations and extension agencies of the land-grant colleges, and officers of the National Association of Land-Grant Colleges. They advised the Authority to produce a *limited amount* of specific types of concentrated phosphate fertilizer with which to carry on tests and experiments under controlled conditions." *Hearings*, Joint Committee to Investigate the TVA, *op. cit.*, p. 122. (Emphasis supplied.)

[23] Vol. II, Pt. I, p. 801.

[24] This study does not purport to present an adequate description or appraisal of the substantive program. These data are introduced only to provide reference points for further discussion of administrative relationships, machinery, and problems.

[25] Admin. Bulletin No. 23, Office of the Gen. Manager, TVA, August 16, 1937 (mim.).

[26] See below, pp. 147 ff.

[27] Admin. Code No. 41-0, Office of the Gen. Manager, TVA, Oct. 7, 1941 (mim.).

[28] See below, pp. 147 ff.

[29] Admin. Bulletin No. 25-1, Office of the Gen. Manager, TVA, July 24, 1939 (mim.).

[30] Of these, 171 were in the seven Valley states, and 30 in twenty non-Valley states.

[31] Admin. Bulletin No. 43-1, Dec. 10, 1940, Office of the General Manager, TVA (mim.).

[32] See below, p. 230.

[33] Of the land-grant colleges, Barton and Morgan, *op. cit.*, p. 101, report that ". . . after allowance has been made for such factors as the utilization of the services of white county agents in behalf of Negroes, interracial meetings, and the segregation in research reports of the data for the races, and the more recent attempts in the work of the experiment stations to give attention to the rural problems peculiar to the Negro, there is still evidence that in research and extension there has been discrimination against the Negro. This appears to be more marked in extension than in research. Experience under the several land-grant acts points toward the need of a Federal safeguard if discrimination on racial grounds is to be prevented." In 1936, J. Max Bond, then responsible for Negro personnel work in the TVA, called attention to the opportunity for bettering the relative position of Negroes in the extension service, without favorable response. The issue was also raised by the National Association for the Advancement of Colored Peoples at the time of the Congressional Investigation of TVA in 1938. See *Hearings*, Joint Committee to Investigate the TVA (*op. cit.*, in n. 11 above), pp. 2383 ff. That the department is conscious of unfavorable criticism is

evident from the suggestion by a TVA agriculturist that work of a white assistant county agent in a small Negro community be given a special publicity "build-up."

[34] See below, pp. 164 ff.

[35] But see above, p. 105.

[36] See p. 151.

[37] As TVA phosphate production increased, the problem of distribution through new mechanisms, beyond the test-demonstration and other government programs, was posed. The result was an important victory for the supporters of farmer coöperatives. In 1944, the TVA publicly affirmed its support of coöperatives as having an important role in assuring an adequate national supply of fertilizers. See the statement of the Board of Directors, "Mineral Fertilizers and the Nation's Security," December 1, 1944.

CHAPTER IV

TVA AND THE FARM LEADERSHIP
(Continued)

In analyzing the relations between TVA and the extension service, the significant question for you to ask is whether the tail has not begun to wag the dog.

A TVA OFFICIAL (1943)

BEFORE SETTING FORTH details of the working relationship between the TVA and the land-grant colleges, a brief review of some important considerations with regard to the agricultural extension services is required.[1] While the experiment stations of the colleges are also involved, the extension services have a central role, accounting for the bulk of personnel and funds assigned to the TVA fertilizer and readjustment programs.

CHARACTER OF THE EXTENSION SERVICE

The network of land-grant colleges which now covers the United States originated in the Morrill Act of 1862, which provided for a grant of 30,000 acres of land or its equivalent in scrip to the states for each representative and senator in Congress. This established an endowment fund, the income from which could be used for the support of agricultural colleges. This income, which varied from substantial to negligible, was later increased by grants under the Second Morrill Act, passed in 1890, and subsequent agricultural legislation. The 1890 statute prohibited support under its provisions for colleges whose admission requirements made distinctions on the basis of race or color, but the establishment of separate colleges for white and colored students was acceptable. In all, sixty-nine land-grant colleges and universities have been established, including seventeen Negro institutions.

In 1887 Congress passed the Hatch Act, which provided for federal support of agricultural experiment stations attached to the land-grant colleges, on the basis of annual grants. An Office of Experiment Stations in the U. S. Department of Agriculture has exercised a rather {118} loose supervision, limited largely to passing upon the legality of projects, leaving much to the initiative of the individual stations.

Agricultural extension work dates officially from the Smith-Lever Act of 1914, providing for a coöperative Federal-State Agricultural Extension Service for instruction and practical demonstration to persons not actually resident at an agricultural college. This act laid the basis for the county agent system, with funds contributed by counties, states, and the federal government. Administration of the act was implemented by the "1914 agreement" between the Department of Agriculture and most of the land-grant colleges. The Department agreed to undertake all extension work through the agricultural colleges. Extension agents were to be joint employees of the colleges and the Department. However, the Department did not undertake to approve the selection of agents, and in general its Coöperative Extension office has not exercised a close or detailed supervision. The Department's supervisory unit is a "mere skeleton organization in comparison with the state and county field forces."[2] The presumptively basic weapon of the Department, the authority to withhold funds, has been used only once—in the case of Puerto Rico,—in addition to a threat against Mississippi in 1930. In general, the state organizations follow their own programs, receiving grants-in-aid from the federal government, with little more than "advice and persuasion" as day-to-day administrative control from Washington. Doubtless the existence of latent broad powers in the federal government—which could be exercised if the situation warranted the risk of political crisis—induces a measure of voluntary restraint against gross inadequacies.

The relation of the TVA to the state extension services follows the pattern established by the Smith-Lever system. It is therefore necessary to refer to certain characteristics of that pattern which have evolved over the years. These characteristics include (1) indications that the county agent is committed to "court house politics" on the county level; (2) the relation of the extension services to a private quasi-political organization; (3) a tendency of the extension agents to deal primarily with more prosperous elements of the farm population; (4) the emergence of new administrative duties of county agents, associated with changing organizational objectives of the extension service.

1. While there is some tendency for financial support of county agents to be restricted to state and federal sources, the traditional pattern of some contribution by the county government prevails. As a consequence, {119} the county agent tends to be continuously involved in factional controversies at the local level and especially to devote a considerable portion of his attention and effort to insuring the continued appropriations of county funds for the extension office. Even where the contribution of the county is relatively small, the extension service of the college will not make an appointment which meets opposition from the county leadership. In the poorer counties of the South, the county agent looms rather large in the local machinery, and his need for political accommodation is correspondingly great. The need to render special favors to potent individuals or organizations cannot readily be ignored by an agent who wishes to hold his job.[3] This situation is generally recognized by extension officials in the TVA area and accounts for some of the reluctance of assistant agents on the TVA program to be promoted to county agent.[4] The special relation of the

county agent to local government emphasizes the grass-roots character of the instrumentality, but at the same time it constitutes a force which tends to shape the TVA program along special lines governed by the preëxisting pattern of leadership and control in the local area.

2. Local farm bureaus have been associated with the extension system since its inception.[5] County associations of farmers were strongly supported and promoted by extension leaders and county agents as useful vehicles for implementing the rural education program. At first, these groups acted as the arms of the extension service, but as state and national federations were organized, legislative matters came to have a predominant place in the attention of the farm bureau leadership. By the time of the organization of the American Farm Bureau Federation in 1920, the political orientation was fixed, so that what began as an educational device became an organized pressure group. As a national machinery was established, and a corps of organizers sent out into the farm areas, the role of the county agent remained central, but he became increasingly dependent upon the farm bureau organization and consequently was often committed to the policies of the national leadership. It is true that local farm bureau organizations do not always or necessarily agree with or carry out the policies of the Washington office of the American Farm Bureau Federation, but their dues contributions support the national organization. In a few states the county farm bureau still has legal status as the agency coöperating with the {120} extension service, which has prompted rival organizations to protest.[6] In the Southern states, the farm bureaus are not as strong as in, for example, Iowa and Illinois, so that the local organizations exercise rather less control over the extension service, and as Baker suggests,[7] the relation of dependence may be reversed. In Tennessee the state director of agricultural extension is a member of the board of directors of the Tennessee Farm Bureau Federation. This is also true in North Carolina and other states in the region. In Alabama, in August, 1942, it was observed that county agents had posted recruiting posters for the farm bureau, which stressed that the American Farm Bureau Federation had procured 110 per cent of parity for the farmers. This was at a time in which the national organization was deeply involved in lobbying activities at the capital.

The Department of Agriculture has recognized the anomalous implications of the extension service-farm bureau relationship. In 1921, a memorandum of understanding between the Department and the American Farm Bureau Federation attempted to distinguish between the farm bureau at the county level, which might legitimately contribute funds to extension service work, and the state and national federations (American Farm Bureau Federation), with which there were to be no official relationships. The agreement provided that:

> The county agents will aid farming people in a broad way with reference to problems of production, marketing and formation of farm bureaus and other coöperative organizations, but will not themselves organize

farm bureaus or similar organizations, conduct membership campaigns, solicit memberships, receive dues, handle farm bureau funds, edit and manage the farm bureau publications, manage the business of the farm bureau, engage in commercial activities or take part in other farm bureau activities which are outside their duties as extension agents. {121}

and further:

The county farm bureaus have their State and national (American) Farm Bureau Federations, which are working on economic and legislative matters and are also promoting the extension service and agricultural education and research. These federations are, however, not directly connected with the extension service and do not enter into coöperative agreements with the State colleges and the Department of Agriculture involving the use of federation funds and the employment of extension agents, and the college and the Department are not responsible for the activities of the farm bureau federations.[8]

In fact, however, these rules have not been generally effective. Twenty years later, it was necessary for the Secretary of Agriculture to issue the following statement:

Recently, reports have reached the Department of Agriculture that officers or employees of the Department have participated actively in meetings and in other activities concerned with the establishment of general farm organizations, or with recruiting members for existing farm organizations.

It has long been the established policy of the Department that its officers and employees shall refrain from taking any part in activities of this type. This is a necessary corollary of the equally long-established policy of the Department that it shall deal fairly with all farm organizations and deal with each upon the same basis.

As a continuation of this policy, it should be understood by all officers and employees of the Department that it is not permissible for any of them to—

1. Participate in establishing any general farm organization.
2. Act as organizer for any such general farm organization, or hold any other office therein.
3. Act as financial or business agent for any general farm organization.
4. Participate in any way in any membership campaign or other activity designed to recruit members for any such organization.

The phrase "general farm organization" used in this memorandum is intended to refer to such national, regional or State farm organizations as, among others, The National Grange, The American Farm Bureau Federation, the Farmers' Union. The Farmers' Holiday Association, The Farmers' Equity League, the Missouri Farmers' Association, and their regional, State and local constituent groups.[9]

This memorandum has apparently not been enforced with respect to the extension service, perhaps because extension officials in the field think of themselves as state rather than federal employees, and because of the long-standing pattern of loose control from Washington.

3. The tendency of the county agent to deal with relatively more prosperous farmers has received considerable comment and some study.[10] {122} Attention to this problem has been increased with the intervention of government agencies, such as the Farm Security Administration, whose work was specifically directed toward the lower third, and the rise of farm organizations such as the Southern Tenant Farmers Union[11] and the National Farmers' Union, which have strongly criticized both the extension service and farm bureau system. In the Southern states the situation is aggravated by the neglect of Negro extension needs, but basically it appears that the nature of the extension service tools, rather than any "Bourbon" attitude, sets the pattern. Extension leaders often admit that their organization cannot reach all the farmers, but they point out that their traditional function has been education. A county agent must work with those who possess the means for undertaking new agricultural methods, and no funds or other material aids have been available for those who could not do so. Consequently, it was natural that the county agent should tend to ignore those who did not have the means to participate, and that he should build up his clientele primarily among those who did have some resources. Home demonstration agents have faced the same problem, since poorer families could not afford to put advice for home improvement into action. As educators, {123} the agents have been under pressure to make a good showing so that they often turned to the more well-to-do farm families who might be used to advantage in reporting accomplishments.

Extension service officials in the South tend to accept the prevailing code with respect to farm tenancy. Thus one farm management specialist in Georgia pointed out that the basis for criticism of the extension service was the tenancy situation. But in extenuation he said, "The extension service must deal with the landlords, but in this way we do reach the 'croppers'." The proportion of farm tenancy is very high, so that the extension service may, in this state, deal directly with only the upper third of the farmers. With tenants having little or no material means of their own, and being highly mobile, it is not surprising that the county agents should restrict their attention to the more substantial farming population, which can participate in a long-time educational program.

4. The role of the county agent as an itinerant teacher, an outpost of the agricultural college, has been subject to considerable changes. Even during the first world war, the county agent played an administrative role in attempting to push a national program of agricultural production. The slow method of demonstration gave way to a series of campaigns, organized around concrete objectives such as crop quotas.

In addition, problems of farm labor and food conservation were added to the wartime functions of the local agents, to say nothing of war bond drives and miscellaneous services. However, the basic orientation of extension as an educational rather than an action organization persisted. Yet the administrative potential of a field force thus widely spread over the country could not be overlooked, and when depressed economic conditions brought large-scale government intervention for the benefit of farmers, the educational aspect of the system was subjected to increased attenuation. Some of the stronger colleges resisted the new tendency, but those with little available funds, especially in the South, welcomed the new programs. Perhaps the most important of these was the work of the Agricultural Adjustment Administration, which provided funds for the employment of numerous additional county agents and assistants. The county agent assumed a central administrative role and was, as a rule, deeply involved in many different action programs. Hence the character of the extension service underwent a fundamental change from that which the Smith-Lever Act envisioned. Far from being merely a teacher, devoted, at least in theory, essentially to noncontroversial promotion of better farming methods, the county agent became an executive {124} with commitments and involvements which forced him into a strategic position in the farm leadership of his community.

These characteristics of the extension service have been reviewed only in order to make clear some of the factors which might be expected to shape the relationship of the Tennessee Valley Authority to the state institutions. In the interaction between these organizations, the character of each represents the effective channel—as distinct from the formal relationships—through which the proposals of policy become disposed inaction.

PATTERN OF COÖPERATION

The coöperative arrangements between TVA and an individual state university have usually been embodied in a master contract for a "joint program of agricultural development and watershed protection through improved fertilization," signed by the president of the college or university and the Chairman of TVA. Statements of purpose and policy are laid down, centering around the encouragement of rational farm management practices which would afford maximum benefit to the farmers from the use of new fertilizer products, all within a "regional program of watershed protection and agricultural development by which soil erosion and flood waters may be controlled and soil fertility may be restored and permanently maintained."[12] Provision is made for a broad coöperative program involving soil surveys in Tennessee Valley counties, laboratory and field research to be carried on at agricultural experiment stations, and farm unit and area demonstrations using the facilities of the agricultural extension service. In general, the actual work is carried on by the college organization, the Authority reimbursing the institution for its expenditures on additional personnel and equipment.[13] Facilities and equipment acquired under this arrangement became the property of the Authority. The master contract is implemented by a series of project agreements which are signed by the TVA's Director of Agricultural Relations and the head of the experiment station or extension service within the college.

These formal agreements are made within a context of loose and traditional ties and, as written instruments, have not always been strictly observed or even kept up to date. While some criticism on this score has arisen within the Authority, in fact the ability to conduct the program {125} on an informal basis is deemed an asset. The grant-in-aid procedure is dominant, with the traditional Smith-Lever pattern[14] consciously maintained. In some respects, the states have an even greater advantage in dealing with TVA than they do with the Office of Coöperative Extension in the USDA. Thus the dean of one of the agricultural colleges in the Valley pointed out that the colleges are not limited to a rigid budget in respect to grants from the TVA, but can return to the Authority for interim revisions of project agreements allowing for increased funds during a given year. The usual role of the TVA agriculturists has been characterized as "passive and advisory," borne out by their persistent emphasis upon the need for programmatic direction to come from the states.

The organizational implications of this pattern of coöperation come in part from the collateral objectives which define the methods available to the Authority for administering its program. These qualifying conditions, self-imposed, concretize the grass-roots policy, presumably offering criteria for measuring the extent to which it has been applied or been successful. Coöperative projects should (1) stimulate the local institutions to an interest of their own in the joint activities, so that they may gradually increase their own financial contributions and ultimately even carry on independently if support by TVA were withdrawn; (2) have a leavening effect on the regular extension service program; and (3) broaden the perspective of personnel in the Authority. These criteria and objectives sustain and guarantee the traditionally loose relationship between the land-grant college organizations and the federal government.[15] Such an interpretation is warranted by the following considerations:

1. The desire to stimulate the local institutions tends to foster relationships which are informal and flexible, thus minimizing businesslike *quid pro quo* attitudes which an emphasis on reciprocal contractual obligations may induce. The colleges may be nursed along and permitted to deviate from formalized procedures, especially if there is the possibility that added responsibilities will be accepted. An unusual amount of consideration and attention of the TVA agricultural staff is devoted to preserving the integrity of the colleges so that they may receive full {126} public credit for participation in the coöperative

program, and particularly to supporting the colleges as indispensable channels to the farm population. At the same time, little systematic attempt is made by the Authority to gauge the extent of contributions by the institutions. While the grass-roots approach has often been discussed and justified, no clear evidence of a significant increase in contributions, especially financial, by the colleges to the joint program has been presented. Even as late as 1943, attempts to make estimates along this line were faced with considerable difficulty. There is, moreover, some doubt as to whether all funds contributed by the colleges are derived from local sources. Grants-in-aid from the U. S. Department of Agriculture are thought of by the colleges as their own funds, and the TVA does not question the sources of funds applied by the institutions to a particular project. Thus there is the possibility that resources used by the institutions may represent actually a federal rather than a state contribution. Lack of clarity on this problem appears to be due to the conviction of the TVA agriculturists that the grass-roots method has an inherent validity which does not require any objective tests. Indeed, the colleges are reluctant to provide information as to their financial contribution to coöperative projects, and one TVA staff member reported that the Director of Agricultural Relations has instructed his department not to request such information from the colleges, "since they don't want to give it."[16]

2. The objective of influencing the regular program of the experiment stations and extension services involves the corollary procedure of integrating the TVA program with that previously developed within the college organizations. The articulation of the TVA and extension services program will be described below.[17] At this point it suffices to suggest that the TVA comes to have a special interest in and commitment to the regular program of the extension services so that, to some extent, the institution can carry out its obligations under the joint program simply by continuing previous services to the farm community. Personnel assigned to TVA work may do other extension work without fear {127} of being questioned by TVA, thus underlining in action the continuity of TVA test-demonstration projects with customary extension procedures. Furthermore, the integration of the two programs vitiates efforts to measure contributions from the institutions, so that relationships are based fundamentally on a faith in the logical indivisibility of extension and TVA activities. The extension service has in fact been influenced by the TVA program, especially in being provided with a means (fertilizer) of material aid to induce farmer participation, and in having a well-defined set of objectives on which effort might be concentrated. However, the very process by which the extension service absorbs TVA fertilizer distribution and testing activities demands that the extension organizations be permitted to be free of over-diligent supervision or inspection from the TVA offices. This has been recognized by the TVA agriculturists who

have consciously refrained, for example, from sending inspection teams into the field to check on the work of the extension service.

3. It is held desirable that TVA personnel will find new and broader perspectives if they consciously attempt to exploit the resources available in the local institutions. This doctrine objectively strengthens the relative weight of the local institutions in their relations with TVA insofar as adaptation to the local organizations (and to the grass-roots theory generally) on the part of TVA personnel elicits official approbation. Far from approaching the local agencies with a "reformer complex," this viewpoint assigns a special value to those agencies and exalts them as a source of education and inspiration. However, in the agricultural program, which is administered by TVA officials who come initially from the colleges, it appears doubtful that any significant education of this sort occurs. Since the TVA agriculturists are themselves products of the extension service system, and among the most vigorous advocates of that system, it is not easy to see how they themselves are broadened by contact with the colleges. However, it is no doubt true that other divisions of the Authority, having different origins, are shaped to some extent by working with the local organizations. In any case, this doctrine of exploiting local resources tends to justify and bolster any influence which the college staffs may have upon TVA personnel.

The strong belief of the TVA agriculturists in the land-grant college system is, as a set of organized sentiments, the fundamental basis of the administrative relationships. This, as has been observed, results from the institutional origins of the leading personnel and from the traditional {128} precedents laid down in the administration of the Smith-Lever Act, as well as from a defensive psychology of representatives of the region. These sentiments are sustained by the grass-roots doctrine, which involves an essentially localist conception of democracy. In addition, an administrative logic of working through the land-grant colleges has been developed, asserting that democratic administrative policy leads inexorably to the pattern of coöperation already laid down. This logic maintains that the TVA fertilizer-testing and soil conservation programs would not have been welcomed by the individual farmer as isolated practices to be appended to his existing plan of work; they require rather a total reorganization of the individual farm plan. "But," runs the argument, "for these conservation and fertilizer practices to be integrated with other farm problems and presented to the farmer as a *unified* program they would require either that the Authority establish an agricultural service which would seek to carry out a complete program of research ranging from purest research to the giving of advice to the farmer on his farm and the farmer's wife in her kitchen, or that the Authority fit its special interests into existing farm programs." If the Authority had elected to create its own agricultural service, it would have had to undertake an educational program at the county and community level which would be in line with the character of the county agent system. Moreover, its field agents would require training, and this would involve the establishment of agricultural colleges, including experiment stations for the controlled testing of farm practices. In the end, therefore, if the TVA had decided to set out on its own, its organization would have been "patterned after the ideals of the land-grant colleges." It therefore makes eminent good sense, in this view, that the TVA should have chosen to work through the existing college system.

It seems clear that this all too innocent formula reflects the fundamental conviction of the TVA agriculturists that there is an identity which unites the TVA program in agriculture and that of the land-grant colleges. Coöperative arrangements with the colleges would inevitably have been undertaken by any federal agency engaged in soil conservation, but the specific pattern of coöperation developed by TVA cannot be adequately understood apart from the ideological bent and institutional commitments of the TVA agriculturists.

The implications of the administrative logic just paraphrased go beyond the relations of the colleges to the TVA. By a similar logic, it would seem that the newer agricultural field agencies, such as Soil {129} Conservation Service and Farm Security Administration, cannot or ought not remain side by side with the colleges, since they tend to converge in function. To the extent, therefore, that there is a struggle among these agencies, the TVA arguments throw the weight of the Authority on the side of the land-grant institutions. To avoid duplication, either the new agencies would have to be absorbed by the colleges, as in the case of the TVA agricultural program, or a drastic reorganization of the extension service would have to be undertaken, divorcing the administration of agricultural action programs from the educational function. The latter alternative would put a wedge between the colleges and their client publics, hobbling their influence. But it is precisely this access to the farmers which is at stake, a prerogative which was amply protected in the pattern of coöperation won by the colleges in their relation to the TVA. Of course, the administrative logic defended in the TVA disregards other agencies that have chosen to build field forces to reach the farmer with a special program. The sheer convergence of formal function (as of education, or of the presentation of an integrated farm program) does not seem to weigh heavily in the minds of those who feel that the traditions and commitments of existing institutions will not permit adequate administration of programs which require a special loyalty, vision, or zeal.

INTEGRATION OF TVA AND EXTENSION-SERVICE PROGRAMS

The joint agricultural program of TVA and the land-grant colleges of the Tennessee Valley includes (1) the controlled testing of plant food products manufactured by the Authority according to standing procedures of the agricultural experiment stations, including work in laboratories, greenhouses, and on field plots; and (2) a program of large-scale testing and demonstration of phosphatic fertilizers on actual farms. The latter is known as the "test-demonstration" program and, carried on through the extension services, represents the major phase of the work.[18] In addition, a program of assistance to farm people and rural communities affected by the flooding of farm lands by TVA reservoirs is also carried on through the local extension-service organizations. Since the activities of the extension service in the joint program represent the TVA channel to the farm people of the Valley area, and constitute overwhelmingly the largest part of the TVA effort in agriculture, they assume a primary importance in the grass-roots approach. {130}

Coöperative agricultural extension work was defined in the Smith-Lever Act of 1914 as "the giving of instruction and practical demonstrations in agriculture and home economics to persons not attending or resident in said colleges in the several communities, and imparting to such persons information on said subjects through field demonstrations, publications and otherwise." The joint TVA–extension-service program is basically a continuation and intensification of agricultural extension work, with, however, some modifications necessitated by the special objectives of distributing TVA fertilizer and carrying on the relocation and readjustment work. The TVA Act authorized the Authority to arrange with farmers and farm organizations for use of new forms of fertilizers under conditions permitting an accurate measure of the economic return they produce. This qualification added a new factor to the extension-service work, for while demonstration had long been a cornerstone of agricultural extension, testing had been left to the controlled, experimental investigations of the stations. Consequently, the TVA adopted the idea of a test-demonstration program, wherein the scientific function of appraising the practical value of concentrated fertilizers would be combined with the educational function of encouraging individual farmers and their neighbors to try the new materials and spread their use.

In 1935, after a year of work devoted to testing Authority fertilizers at the experiment stations, the test-demonstration program was inaugurated. Broadly conceived as a contribution to the total agricultural problem of the region, this program of distributing phosphatic fertilizers was to serve a dual objective: water control on the land in conjunction with the adoption of methods of farming beneficial to the welfare and security of farm people in the area. Phosphates are integral to a system of farming which emphasizes the growing of leguminous forage plants or cover crops. In this way the soil is enriched not only by the application of phosphate and lime but by the fixation of nitrogen from the air. At the same time, the legumes (such as alfalfa, lespedeza, red clover) hold down the soil and prevent the erosion which follows upon the planting of such clean-tilled crops as cotton and corn, especially in hilly regions. Moreover, a shift from a corn-and-cotton economy to one which places a greater emphasis on livestock is encouraged; this has long been an objective of agricultural agencies in the South. Hence the obligations of the TVA to protect its reservoirs and decrease the danger of floods, as well as its responsibilities for fertilizer manufacture and distribution, are brought together in a program readily articulated {131} with the educational objectives and machinery of the agricultural extension services.

The fertilizer distribution program utilizes two basic devices, individual farm test-demonstrations and area test-demonstrations. The work with farms brought into the program through these devices represents the basic and dominant phase of the joint activities. The machinery of the extension services is used exclusively,[19] the TVA reimbursing the colleges for the employment of assistant county agents who, working out of the county extension offices, are assigned to the test-demonstration work. In addition to the assistant county agents, supervisors and other specialists are employed by the colleges with salaries reimbursed by TVA. These officials of the extension service constitute the operating organization which brings the TVA fertilizer and soil conservation program to the rural population.

In theory, and to a varying extent in practice, farmers on the program are selected by the communities themselves, through meetings called by the county agent. The distribution of fertilizer is handled by a county soil erosion control association to which bulk consignments are made by the Authority. However, there is no question that the county agent (or the assistant county agent assigned to the program) is the key man. It is "his" program and "his" association, and it is understood in practice that all contact with the farmers and the farm organizations which implement the program will be made through him.[20] The selection of farmers to participate in the program is in practice the responsibility of the extension service official, though this is modified to some extent by the use of community committees and meetings. Sometimes the already established AAA committees were used to inaugurate the test-demonstration program.

The unit test-demonstrators usually enter the program for a period of five years. Each agrees to commit his entire farm to the testing program, involving the reorganization of the farm plan to increase the production of food for the family and feed for the livestock; to readjust land-use practices so as to halt erosion and build up the soil; to use the phosphatic fertilizers provided by the TVA, combined with lime, to increase the acreage of cover crops; to leave check plots untreated for comparative purposes; to pay freight charges on shipment of fertilizers; to keep records and provide access to the farm by extension service personnel and other farmers. These contributions by the {132} farmer are strongly emphasized, in order to dispel the idea that free fertilizer is being distributed.

The area test-demonstrations involve the distribution of fertilizer to an entire watershed community. At the initiative of the county agent, a community organization is formed, and a survey of the area and inventory of its resources is made as a basis for the plan under which each farmer will participate in the soil conservation program, receiving TVA phosphate as a material aid. An attempt is made to hold regular monthly meetings at which farming plans and other community problems are discussed.

In Tennessee, in 1942, there was one farm unit test-demonstration for each fifty farms in the Tennessee Valley counties of the State; area test-demonstrations included about four per cent of all farms. It was calculated that costs to the public per test-demonstration farm (unit and area) amounted to $100 a year, one-half allocated to the cost of the phosphates and the other half to salaries and travel of extension service personnel. The farmers themselves are believed to have invested about an equal amount in lime, terraces, and other improvements.

A basic procedural principle which runs through the entire administration of the test-demonstration program is the goal of integrating that program within, and thereby strengthening, the county agricultural extension program. Each step in that direction, each evidence that the extension service has absorbed the TVA program into its own, is considered an achievement. In the relocation and readjustment program as well, although dealing with problems of an extraordinary character, the ultimate objective is the attainment of a stage which will permit the job to become part of the regular program of county extension work.

The initial phase of aid to reservoir-affected farm families and rural communities is that of relocation. A series of conferences involving the Authority and Extension service officials, extension workers, county program planning committees, and subordinate county relocation committees, down to neighborhood groups meeting with county agents, are held. Detailed surveys of the area are made. Lists of farms for sale or rent are compiled and appraisals made by extension officials. Information and more material aid (such as in moving) are made available until the flooded-out families have removed to their new homes or left the area. Steps in the process of population removal and readjustment have been formulated as:

1. Study and analysis of the social and economic conditions of the region affected by reservoir flooding. {133}
2. An appraisal of all the resources available for readjustment.
3. Formulation of plans for the evacuation of the reservoir area and the readjustment of the displaced families.
4. Visitation of all families living within the reservoir area.
5. The removal of families from the reservoir area.
6. Readjustment of families at their new locations.

During the relocation phase, the Reservoir Property Management Department of the Authority takes an active part, with the extension service the main coöperating agency. Up to April, 1943, thirteen reservoir areas had been evacuated, displacing 9,489 families, in addition to 1,819 families removed from areas of projects then under construction.[21] The Reservoir Property Management Department reported that only five families were evicted by court order, all from the Norris area (the first to be evacuated) where plans had not been put into action before the rising of the lake. Farm ownership has varied from 59 per cent in the Norris area to 7 per cent at Wheeler dam, averaging 35 per cent of the total families displaced. Less than 6 per cent of the families removed up to 1943 were Negroes.

As the process of relocation is completed, the readjustment phase begins. It is here that close integration with the regular county extension program is made possible. Additional assistant county agents and supervisors added to the extension staffs for the readjustment program (with salaries reimbursed by TVA) devote themselves to aiding farmers to achieve a successful adjustment in their new locations. The test-demonstration program, especially the area watershed device, has been used as a vehicle for bringing educational and material aid to these farmers. During this phase, the duration of which is not defined, the test-demonstration program is extended on the basis of the needs of the relocated farmers. Consequently the same articulation with the extension service obtains in the readjustment program as in the test-demonstration work generally. The assistant agents assigned to readjustment work may, in addition, be called on to assist in the administration of the rental of Authority-owned lands through the land-use associations set up in the reservoir areas.

With this brief sketch of the program as background, we may now turn to a consideration of some of the problems raised by the TVA channeling its agricultural activities through the extension service machinery. It will be evident that the discussion above of the character of the extension service[22] underlies the following analysis. {134}

1. *The "testing" dilemma.*—Section 5 (b) of the TVA Act authorizes the Authority "to arrange with farmers and farm organizations for large-scale practical use of the new forms of fertilizers under conditions permitting an accurate measure of the economic return they produce." This authorization is the basis of the test-demonstration program, which in turn represents the core of TVA agricultural activities. Yet, in the administration of the program, there appears to be a constant straining of the actual operations against the official interpretation. The county agent system is not a research organization and its personnel in the counties are not much concerned over problems of scientific appraisal. It is sometimes said that extension personnel tend to be evangelistic rather than scientific in outlook. They conceive of their job as educational, as it traditionally has been, and they are more interested in the demonstration than in the testing aspect of the fertilizer program. Once controlled tests are made at the experiment stations, these agents conceive of the value of the concentrated phosphatic fertilizers as established, the job being to prove to the farmers in the local

areas that they are indeed useful and profitable. In general, the extension agents use the TVA fertilizer as a tool in developing the general agricultural extension program in the county. This, being in line with the conscious policy of integrating the program with that of the extension service, is approved by the TVA. But at the same time the possibility of isolating the effect of using TVA phosphates is reduced, thus undermining the testing function.

The extension service program is broadly educational, related to the general development and welfare of the agricultural community and its resources. The statutory authority of the TVA, however, is confined to the limited problem of fertilizer manufacture, testing, and distribution. But not only is the TVA organization actually interested in promoting the same program as that of the extension service, but it is often said by the Authority's agriculturists that "the TVA has no program of its own in agriculture." The distribution of TVA fertilizers becomes in fact a material aid to be used in advancing the extension service program. As a result, the TVA personnel tend to be unclear and sometimes inconsistent in interpreting their program. The possibilities of attaining an accurate measure of the results of using phosphates on a farm will be dissipated as more factors are added which affect the economic return of the farm unit. But not only does the extension service, through mustering educational resources in helping the farmer, add to the factors, but there are also attempts with TVA {135} support to include the test-demonstration farms in a rural electrification program and to concentrate farm forestry work on the same farms. At the same time, it is insisted that the farmers are testing the value of TVA phosphates.

This difficulty is recognized by the Agricultural Relations personnel, and there has been some suggestion that the testing and demonstration functions should be separated, frankly recognizing that the county agent staff is concerned exclusively with the latter. Moreover, there is some tendency to move to an emphasis on the area demonstration device, which is much less closely specified or controlled than the individual farm demonstration, and is justified partly on a somewhat different basis. The mass distribution of fertilizers to an entire community is viewed as a contribution to TVA's flood-control and dam-protection responsibilities (holding down the soil and storing water in the soil through cover crops) rather than as a use of the narrow fertilizer testing authority. That the idea of testing is partly ideological is shown by the emphases upon it when convenient, and its discard when inconvenient—in terms of the real objectives of the program. Thus fertilizer has been refused to a group-farm project of the Farm Security Administration on the ground that it was an unrepresentative type of farming, but fertilizer has been used on equally unrepresentative types of farms when these have been extension service projects.

2. *The representative farm.*—The contracts with the colleges specify that the test-demonstration program shall be based on farms of representative size, soil variety, and type of farming. However, here too some inconsistency seems to be induced by the relation to the extension service, though there has been considerable effort to make it a real criterion in selection of farms. When, early in the program, one member of the Authority staff—not an agriculturist—raised the question of using the test-demonstration program to reach progressively lower economic layers of farm families, the Director of Agricultural Relations countered by pointing out that it was necessary to get "a representative sample" of farms in the area in order to fulfill the testing function. During the same period, however, county agents in the field were recommending farmers who had worked closely with them for the "free phosphate," usually the more substantial farmers in the community. This tendency would appear to have been a predictable consequence of channeling the program through the extension service. The Agricultural Relations Department has had to use pressure in order to get enough small farms on the program to avoid criticism, and even this has not {136} been accomplished. The problem posed by the attitudes and traditions of extension service personnel is highlighted by the following statement, made in 1935, by a supervisor on the test-demonstration program.

> In spite of the possibility of being thought obstinate, I am going to repeat that I think you folks there are entirely wrong in your idea that small farms should be included in the demonstrations in the hilly grazing counties of Virginia. We have been telling these farmers for years that livestock farming was the only type of farming that could be expected to permanently succeed in this section and have further deluged them with bulletins, farm management records, and speeches in which they were told that farms of less than 160 acres in livestock sections could rarely be operated efficiently.... I think if we try to use small farms for demonstration in this area. . . our books will always show a loss unless we defeat the purposes of soil conservation by encouraging the growing of crops on these farms that would necessitate frequent plowing and cultivation. . .
>
> The farmers who live on the small farms in this area are mostly ignorant and inefficient and are not capable of conducting a demonstration for the benefit of their neighbors. Just yesterday I was told that . . . the small farmers were suspicious of the demonstration idea and that it would probably take considerable educational work to get even a few of them interested....

With the TVA wedded to the extension service machinery, it is not surprising that the program should have been weighted by farmers selected in terms of local commitments of the county agents. The statement of another field agent describes the typical extension service problem in which the available tools determine the extent of participation; the statement is in explanation of why the average size of farms on the test-demonstration program in 1936 in his area was overly large:

> Harlan is a mountain county. It is in the heart of the greatest coal mining industry in our state. The hills are high, and the valleys generally not more than 100 feet across, making the choice of land for fertilizer demonstration purposes quite different from that in the broad valley of the TVA drainage basin. This, to my mind, is an instance where theory and practice lock horns, resulting in the practical side winning, because of the fact that the men with the larger holdings have become demonstrators because it is more feasible for them to sow a cover crop of small grain, and show an economic return. These men are most apt to have teams and tools to prepare the ground; they are most apt to have the cash to pay the freight on the fertilizer; they are most apt to be able to buy grass and legume seed to sow in the small grain next spring, in that general territory. The man who lives at the mouth of the creek, where generally the bottom land is widest, is usually the man who dominates the thinking of nearly all the people who live up the hollow behind him. His word alone will usually determine whether or not the other folk take up a new practice....

Even a sincere desire to work with small farms cannot prevail against factors which determine, for extension service personnel, the possibilities of doing a job which will produce favorable results. As a consequence, {137} the TVA program tends to be shaped by the same factors. A program which emphasizes the increase of pasture lands and livestock finds it difficult to avoid concentration on those farms which can make a measurable contribution.

In 1947, TVA reported test-demonstration farms distributed with respect to size:[23]

Under 50 acres........................ 12.7 per cent
50-99 acres............................ 26.4 per cent
100-179 acres.......................... 30.5 per cent
180-379 acres.......................... 23.5 per cent
Over 380 acres......................... 7.1 per cent

Although comparisons with census figures are not conclusive because of the prevalence of part-time farming among those holding small acreages, this distribution seems to indicate a somewhat disproportionate participation of moderately large farms, given the nature of farming in the Tennessee Valley area. Some indication of relative proportions which would be acceptable in a representative sample is derived from the specifications originally devised for selecting a sample of unit test-demonstration farms for intensive analysis in connection with a TVA and Tennessee–extension service progress report. Out of 61 sample farms sought, it was thought desirable to have 27 under 50 acres, 17 from 50–100 acres, and 17 over 100 acres. The actual sample finally selected included only 6 farms below 50 acres in size, because not enough small farms on the program met the other specifications for the sample, including length of time as participants.[24]

In general, the unit test-demonstration farms, representing the core of the program until at least 1940, have been disproportionately large. Thus, in the Tennessee 50-farm study just mentioned, the sample farms "averaged nearly 60 per cent larger in total acreages than the average East Tennessee farm (including part-time farmers) and had nearly twice as many acres of cleared land." This sample, moreover, was described as more nearly representative of farming in the area than the entire group of unit test-demonstration farms which had been on the program more than six years. In 1940, the farm-unit test-demonstrations in the Tennessee Valley area of Tennessee averaged 213 acres, while {138} according to the University of Tennessee the average size of farms in that area was 79 acres in 1939. However, the program of area test-demonstrations, involving the distribution of fertilizer to farmers in an entire watershed community, has operated to decrease the average size of farms receiving TVA fertilizers. It appears that where county agents have exercised discretion in selection, as in the farm-unit participants, they have leaned toward the larger farms. When, however, the community is taken as the unit of action, the program reaches broader layers.

There is no implication here that the TVA agriculturists wanted unrepresentative participation; on the contrary, they were concerned about the problem and applied pressure upon extension personnel to rectify the situation. As in extension work generally, the means available, and the objectives of the program tend to select out those farmers most capable of contributing to and benefiting from it. Moreover, this pattern becomes institutionalized, including the creation of a supporting set of attitudes, within the extension service. Consequently, the channeling of the TVA program through the extension service created an initial presumption that this pattern would be continued in the test-demonstration work.

3. *Farm tenancy.*—The educational activities of the extension services tend to be largely oriented toward farm owners. The man who owns his farm can plan his farm management and has a permanent perspective. These conditions make him a desirable client for an educational organization. In the TVA test-demonstration program, the same factors hold, and are emphasized by the established extension service pattern through which it is channeled. A farm-unit participant agrees to enter the program for a five-year period, and this alone must eliminate many tenants. The approach in extension service and TVA has been to deal with tenancy through the farm owners, considering sharecroppers as part of the larger farm unit. While many tenants are doubtless receiving TVA fertilizer through the area demonstration work, no special attention is paid to the tenancy problem within the framework of the test-demonstration program.[25] In fact, there is evi-

dence that the Agricultural Relations Department does not consider high or increasing rates of tenancy a problem.

In the relocation and readjustment program, the problem of tenancy cannot be ignored since the majority of farmers removed from the reservoir areas have been tenants (see table 7). Yet the TVA has taken the {139} view that only the extension services have facilities for dealing with the readjustment problem. The logic runs as follows:

> The TVA has no facility for advising land owners who have sold their property to the TVA in the matter of purchasing other farm lands. The Authority is not in a position to know what lands are available and whether or not the price for which the lands are offered is reasonable. The Authority is furthermore not in a position to advise land owners who have sold property to the Authority whether or not they should reinvest in land in their county of present residence or whether they should invest in lands further afield. The Authority furthermore has no facility available for agricultural advice, agricultural assistance, or agricultural education for both land owners and tenants displaced by its land purchase program. It furthermore does not organize 4-H clubs and does not offer any other benefits that have been performed by the Extension Service.

While TVA has given some material assistance in the actual process of relocation of tenant families, the readjustment phase has been undertaken by the extension services, with funds supplied by the Authority. In general, the approach to the tenant problem is summed up in the following extract from a project plan for relocation of families displaced at two of the Authority dams:

> A special effort will be made to find locations for landowners who are willing to provide locations for tenants and croppers now residing on their property or on neighboring property and wishing to continue the personal tie now existing between them and the tenant or cropper. The Authority will give preference of employment in areas surrounding the dams to tenants, croppers and small landowners recommended by the Extension Service as most deserving of an opportunity for accumulating capital to be used in meeting their relocation problems. Such an arrangement will be kept confidential.

There has been some indication that landlords in areas to which reservoir-displaced families have moved have been able to weed out their own poorer tenants and replace them with better ones from the reservoir area, adding to the problem of the most depressed layers of the rural population. In situations of this sort, the relocation of tenants from the reservoir sites might be accomplished, and yet indirect consequences for the rural community might remain uncontrolled.

Neither the TVA nor the extension service appears to have handled the readjustment program in relation to general population problems in the region. In 1934, it was suggested within the Authority that a corporation be organized to purchase from holders of foreclosed farms options on large tracts of fertile land which could then be sold to displaced farmers on a credit basis. Nothing was done, however. In 1940, a sociological study group reported, following a brief inspection of the program, their "dismay at the insulation of the relocation program from {140} that of the Resettlement Administration—insulation that prevented even referral of sharecroppers and tenants who were forced from reservoir property, while extensive services and follow-up were given to farm owners who were forced to leave." This insulation, while not complete, arose in part from the organizational hostility of the extension services to what they considered rival farm agencies. The same group made the following statement:

> In our observations, we discovered a tendency to depreciate the few belated sociological studies that were conducted. Some of the studies are not available for groups such as ours or those at state universities (e.g., studies of relocated families), and the TVA staff itself is unfamiliar with most of these sociological studies—how different from their knowledge of researches in other fields I We could not discover an instance in which any of the results were utilized (e.g., a study of the Governmental Research Division revealed that 95 per cent of the people in Wheeler Reservoir area were tenants, yet, so far as we could discover, relocation assistance and follow-up concerned itself only with those farms which were purchased for reservoir property).

On the whole, it appears that the acquisition of TVA agricultural personnel from the extension service and the basic commitment of TVA to carry on the program through the extension service seriously inhibited the Authority's capacity to deal with the special problems of farm tenancy.

While the relationship between TVA and the extension services naturally implies that most influence should flow from the latter to the former, the weight of TVA-subsidized activities and personnel has not been unnoticed in the work at the county level. Work on the test-demonstration farms, with county associations of farmers and community organizations, provides a definite framework for the extension program which otherwise often tends to be somewhat diffuse. There is, therefore, a noticeable tendency for the center of work at the county level to shift to the TVA assistant county agent. This has resulted in some dissatisfaction among the men, who sometimes feel that they are doing "the really important job in the county" and yet are held to the status of assistant agents. This subordinate status is assured, however, so long as the men assigned to the TVA program are placed within the extension service chain of command. Moreover, the assistant agents are not dependent upon county funds for their salaries, so that they tend to be independent of courthouse politics. As a result, extension officials find that some assistant agents are reluctant to take over the position of county agent. Though the character of their work might well entitle them to a higher rank and salary, the pay of a TVA-subsidized assistant agent is deliberately kept below that of the county agent because, as {141} one extension official said, "It's hard enough to get them to take over a county as it is." The salaries of the assistant agents are determined by the colleges, and it is the internal administrative problem of the extension service staff which may decide the amount paid, rather than TVA.

The TVA-subsidized personnel, including supervisors, specialists, and assistant agents, sometimes exhibits a tendency to become a "little TVA" within the college staff. This is a normal consequence of having a definite set of activities and being paid out of special funds. Thus one supervisor in the field, though officially unconnected with TVA, considered it expedient to know the TVA staff at Knoxville and especially to be acquainted with the real center of authority within that staff. At times, the college officials must make positive efforts to pull the staff together within the single hierarchy. However, this tendency toward a split in the extension service organizations is remarkably small, due in part to the conscious effort made by the TVA agriculturists to maintain the integrity of the extension service organization and halt all attempts to by-pass the formal lines of authority. In 1942, the TVA held a training conference for the assistant agents on the joint program, but some doubts were expressed by the Director of Agricultural Relations whether this would not make the men think of themselves as TVA personnel, an eventuality which would be contrary to the policy of both TVA and the extension services.

The TVA agricultural program has been successfully integrated with that of the extension service. This has been accomplished by beginning with the assumption that (1) the TVA could not legitimately elaborate a program which would be different from that advocated by the grass-roots institutions of the area; and (2) the already established extension services represent the most logical and technically most adequate avenue to the farm population and its problems.

RELATIONS TO THE FARM BUREAU

The traditional partnership of the extension service and elements of the American Farm Bureau Federation[26] raises the question of the extent to which a similar relationship has prevailed between the Farm Bureau and the TVA. Theoretically, and in the light of the above discussion, it might be expected that the TVA–extension service and extension service–Farm Bureau relations would be transitive. On the basis of available evidence, this appears true. The TVA agriculturists have apparently simply carried over from their extension service days an outlook {142} in which close coöperation with this private organization and its state and national leaders is taken for granted.

A contract little known even within TVA, entered into by the Authority with the Tennessee extension service, is the basis of one phase of practical relations between the TVA and the Farm Bureau. This agreement, known as Supplementary Project No. 5, under the master contract between TVA and the University of Tennessee, was signed in 1935 by the Tennessee Director of Extension and the TVA Director of Agricultural Relations and continued in effect from year to year without formal renewal. A "statement of relationships" recalls that the Agricultural Extension Service of the U. S. Department of Agriculture and the national and state Farm Bureau Federations have a history of coöperation for mutual objectives; that the Tennessee extension service has undertaken research and demonstration work on behalf of the TVA; and that the "advice, assistance and coöperation of the Tennessee Farm Bureau Federation will be of great value in the conduct of such demonstrations...." On the basis of this understanding it was stipulated:

> I. That the Agricultural Extension Service shall arrange with the Tennessee Farm Bureau Federation for the assignment of one or more of the principal officers thereof who shall devote his or their full time in furtherance of the purposes of this project.
>
> II. The Agricultural Extension Service of the University of Tennessee shall arrange for the use of the experience, facilities and personnel of the Farm Bureau Federation in the organization of conservation and terracing clubs, the distribution of fertilizer and the conduct of such educational work as will result in a wider diffusion of the mutual aims and objectives of sound agriculture through redirected farm practices in a regional program of watershed protection. It is mutually understood and agreed that all activities of the Tennessee Farm Bureau Federation pursuant to the terms hereof shall be conducted without regard to membership in said organization.
>
> III. The Tennessee Farm Bureau Federation shall interpret for the national organization of the American Farm Bureau Federation, and through it to the National Council of Farm Organizations, the aims and objectives of this coöperative program of readjusted agriculture and watershed protection and shall from time to time indicate to the Authority through the Extension Director the modifications in the plan of procedure recommended by the National organization as being of special significance in the solution of the national problem of land conservation.

This contract was also signed by the President of the Tennessee Farm Bureau Federation as an agreement between that organization and the University of Tennessee. A supplementary memorandum provided for a special field agent, to be designated Director of the Educational Division of the Tennessee Farm Bureau Federation and paid by the Farm Bureau which would then be reimbursed by the extension service "from money appropriated by the TVA for this purpose." Notwithstanding {143} the provision of the last sentence of section II quoted above, it is reported that the agent in question has functioned in fact as an organizer for the Farm Bureau. One TVA official, a participant in the agricultural program, said that he himself had attended a meeting at which the agent spoke for the Farm Bureau, and that in general he felt the arrangement to represent "an unhealthy subsidy of a private organization by federal funds." "However," he continued, "this is not an unusual situation in the relations between extension service and the Farm Bureau. Many of the extension personnel do organization work for the Farm Bureau. In Mississippi, the president of the state Farm Bureau is on the board of trustees of the Mississippi State College and he exerts pressure on the extension service to help in membership drives, which is why there are about 30,000 Farm Bureau members in Mississippi, about 10 per cent of the farmers." The report of this official with regard to the Farm Bureau agent paid out of TVA funds has been confirmed by at least one other individual.

Not only has the TVA continued the day-to-day relations to the local Farm Bureau, which are normal in the extension service, but there is also evidence that there has been close contact between leading agriculturists in the Authority and the national leadership of the American Farm Bureau Federation. These relationships have extended to personal conferences on legislative matters and interagency controversies.

Thus, on June 27, 1941, Mr. J. F. Porter, President of the Tennessee Farm Bureau Federation, wrote to Mr. E. A. O'Neal, President of the American Farm Bureau Federation, in a vein which could leave no doubt concerning the close relation between TVA and this national farm organization. Mr. Porter reported that he had held a conference with Dr. H. A. Morgan on the land-grant college situation. Dr. Morgan voiced his concern regarding attitudes expressed in some quarters toward the land-grant colleges and stated that he would like to coöperate with Mr. O'Neal in the matter. He suggested that President Roosevelt should be informed concerning the situation and that TVA might well intercede if the large farm organizations would ask TVA to make a study of the land-grant college controversy. TVA would then take up the matter with the President. Mr. Porter indicated that this procedure would not take much time, since TVA was already well informed. He also discussed the question of TVA's jurisdiction over the Cumberland Valley area, indicating that he had assured Dr. Morgan of the Farm Bureau's support.

Following these discussions, a meeting was held late in July, 1941 at Muscle Shoals, Alabama, attended by Dr. H. A. Morgan, Mr. J. C. {144} McAmis, Mr. Neil Bass, Mr. Porter, and Mr. O'Neal, at which the implications of interagency controversy for the role of the TVA were discussed.

This evidence explicitly supports the interpretation that relations with the Farm Bureau have been national as well as local and have involved matters charged with political implication. It appears that Dr. H. A. Morgan, in effect a representative of the agriculturists on the TVA Board, has been ready to intercede with the President of the United

States on behalf of the land-grant college system. At the same time, the President of the American Farm Bureau Federation, not ungrateful for support of the institutional basis of his own organization, lends his good offices in matters of legislative interest to the TVA. It will be recalled (p. 78 above) that in 1941 the Department of Agriculture was among those who opposed a bill for the inclusion of the Cumberland River basin within the jurisdiction of the Authority.

The close relation between the TVA and the Farm Bureau reinforces the initial commitment to the extension service system. Alternatively, it is evidence that the relationship to the extension service is something more than a formal coöperative arrangement for stated objectives. The concrete operation of the grass-roots policy has, in this respect, led to consequences unlooked for by the founders of the Authority. One of the most important of these founders was Senator George W. Norris, whose views on TVA and the Farm Bureau are, consequently, of interest:

> Senator NORRIS: I did not intend to mention it, but since the Judge has brought it out, I will say that I have had some correspondence with one of the farm organizations in criticism [of TVA's relation to the Farm Bureau]. I think it is proper for me to say that, because I have been connected with the TVA, or with what the TVA has been doing, and I was unaware that they were doing anything that I thought was wrong about it. But I took the matter up...with the idea of ironing out what I thought was just a simple little...case of jealousy, that is probably natural, that is almost human—they thought the Farm Bureau was getting too much attention, and they wanted some of it... I tried to convince them, and I hope I succeeded, that there was no possible ground for complaint that the TVA was picking out any farm organization; that they were glad to coöperate with all sorts of organizations, and I called their attention to the law, which provides what they should do, and that they had not done anything with the Farm Bureau that they would not do with every other organization.[27]

Senator Norris does not appear to have been fully informed in this matter. It was this same senator who, speaking of the Farm Bureau lobby in 1928, predicted "...the time will come when the rank and file of American {145} farmers will begin to realize by whom they are being deceived here in Washington."[28]

Apart from the general effects of the relationship with the extension service, there is little indication of any special influence by the Farm Bureau on the TVA program. However, it is reported that the Farm Bureau in Tennessee has undermined efforts to organize farm coöperatives on the theory that a network of such coöperatives might become sufficiently strong to be independent of the extension service and the Farm Bureau. A proposal to organize an association of coöperatives is said to have been quashed under Farm Bureau pressure. In general, there is said to be passive resistance on the part of the Farm Bureau and the extension service to the organization of coöperatives. Since the TVA works only through the extension service in such matters, any orientation of the Authority in that direction would be inhibited.

AN ADMINISTRATIVE CONSTITUENCY

The foregoing analysis of the relationship between the TVA and the farm leadership in the Valley area may be summarized in the concept of the administrative constituency. The idea of coöperation or working with and through local institutions is qualified in its concrete reference by the sociological dimension of administrative behavior. This dimension recognizes the force of informal goals and institutional commitments, and to that extent serves to make explicit some well understood but seldom formulated bases of executive action. A constituency is a group, formally outside a given organization, to which the latter (or an element within it) has a special commitment. A relation of mutual dependence develops, so that the agent organization must defend its constituency and conversely. This relation gains strength and definition as precedents are established in behavior and in doctrine, and especially as the constituency itself attains organized form. A group which finds its coherence in common interest, but remains unorganized, may enforce its demands in subtle ways, but a leadership and a machinery serve to mobilize its resources. At the same time, however, this machinery may become separated from its popular base and itself become the effective constituency.

The idea of an administrative constituency, however it may be phrased, is familiar to students of public administration,[29] and in general {146} may be thought of as a normal mechanism of social control whereby formal organizations are made responsive to relevant publics. Or, put in other terms, the creation of the constituency relationship is a form of coöptation, the informal involvement of local elements in the process of policy determination. One TVA staff member felt that the TVA's agricultural constituency operated as "one element in the general trend in the Authority toward conservatism and adjustment to Valley institutions." It is this adjustment which is one of the significant implications of the process of coöptation in administration.

The constituency relation varies widely in source, in intensity, and in meaning. One form may arise simply out of the need felt by an organization, public or private, to defend its continuing working relations with an outside group or agency. Interest may focus only accidentally upon the particular outside group involved, for it is the smooth avenue of operation which is being defended, and will not be readily jeopardized in the interests of a program in which the organization may have no great stake. It is often inconvenient and difficult to alter established procedures, so that a given form of coöperation may be defended in order to preserve the integrity and equilibrium of operations. This situation is analytically quite different from that in which the outside organization is defended for its own sake, as occurs in the relation between the TVA Agricultural Relations Department and the land-grant colleges. There is also a distinction to be drawn between short-run pressures upon an agency and the long-run strategy by which a measure of continuing control is achieved. In connection with the administrative relations analyzed above, the land-grant colleges have sought and gained a significant measure of influence upon the TVA; moreover, the TVA agriculturists do defend the colleges as valuable, and even indispensable, in themselves. That a constituency relation in this advanced form exists is well known in the higher circles of TVA, and is the subject of much comment.

The significance of a constituency relation depends in part upon the fact that the character of the constituency will tend to define and shape {147} the character of the agency. This may involve the recruitment of personnel from the ranks of the constituency and, in an extreme form, the assumption by elements within the agency of a leadership status with respect to the constituency. The representatives of the constituency within the agency then come to define their role as one of leading a broad struggle for the furtherance of the interests of the constituency, and sometimes may be more conscious of those interests than members of the outside group themselves. Where the outside group is not the formal source of public policy, the constituency relation may remain more or less covert, and its representatives may find it necessary to devise and rely upon some doctrine or ideology to cover and defend the real relationships. An adequate comprehension of the full meaning of the grass-roots policy does not appear to be possible apart from some such principle as this.

In the relation of TVA to the land-grant colleges, the mechanics of representation include attempts by the Agricultural Relations Department of TVA to (1) channel all activities which may possibly be interpreted as within their subject-matter field through the land-grant colleges; (2) make itself the sole point of contact by the Authority with the institutions; (3) actively oppose all encroachments on the prerogatives of the colleges; and (4) further the policies of the colleges within the Authority, as opportunity may arise.

1. The TVA agriculturists have tried to enlarge the scope of the activities that are carried on through the colleges, to include responsibilities of other subject-matter departments within the Authority. In this respect the approach of the agriculturists has been not toward local agencies in general, but toward the extension services in particular, even when that may mean making a choice against possible alternatives among the state agencies. The history of the TVA Forestry Relations Department points to a different attitude toward local institutions from that utilized by the TVA agriculturists, and provides concrete evidence of the attempt to swing the foresters into line with the organizational methods of the agricultural group.

The TVA forestry division was originally organized as a part of the A. E. Morgan wing of the Authority, under the leadership of E. C. M. Richards. An aggressive program of reforestation and watershed and reservoir protection was conducted primarily on a direct-action basis, using camps of the Civilian Conservation Corps assigned to the TVA by the Forest Service, and not hesitating to deal directly with local land-owners in the distribution of tree seedlings. Work was undertaken {148} with various agencies, including the State Foresters and the U. S. Forest Service, but on the whole the TVA forestry group felt that it had a program of its own, which could not be effectively delegated. However, a convergence of interest with the extension services was soon evident, and this became the basis of a protracted controversy with the Agricultural Relations Department of the Authority. This convergence referred to the joint interest of TVA and the extension service in farm forestry work and in education for forest-fire control, both deemed by the extension service to be within the province of the county agent organization. The professional foresters under Richards, however, did not believe that the extension service could adequately carry out the program. At a meeting early in TVA history, which included H. A. Morgan, J. C. McAmis (Director of Agricultural Relations), and Richards, the latter outlined the basis on which he thought the forestry division might enter farm forestry work. This would involve having the forestry division itself hire and direct the activities of men who would do the work, though coöperating closely with the county agents. The agriculturists, however, asserted that such a procedure would encroach on the field of the extension service, and requested that Richards approve turning over TVA funds to the extension services for farm forestry work, allowing them to hire and direct the activities of the men. Richards refused to do this, and the matter lay dormant for some time afterward.

The attitude of the TVA Chief Forester, and the intensity of the controversy, may be judged from the forceful statement made by Richards in 1938, when the conflict within the TVA Board was under public investigation. This statement is especially important in the light of the entire analysis thus far presented:

> There are decided differences of opinion within the TVA as to just what the TVA is or should be. Three views exist and fight persistently for acceptance:
>
> 1. That TVA is a federally owned and operated electric power utility corporation. All else is quite secondary and is only permitted to exist because (a) it, like dam construction, must go on if electric power is to be generated; or (b) the other activities are so generally popular and worthwhile that it takes more courage on the part of the advocates of this "power concept" openly to throw it overboard, than to pay lip service to it and quietly work under cover to hamstring it and keep it insignificant.

2. That the TVA is an adjunct to local agencies, such as the land-grant colleges and extension services of the seven valley states. Except for dam construction and similar work, which manifestly is quite impossible for the local and state agencies to perform, the advocates of this concept believe that the Authority itself has nothing in the way of a job to do; that it should merely get money from Congress, pass it out to local and state agencies, and let them do everything in their own way, with personnel of their own choosing. Knowing little or nothing of either forestry or wildlife, {149} this group holds firmly to early pioneer ideas of land ownership, land use, and conservation. It thinks of forests as small scattered farm woodland tracts. It thinks of foresters as extension workers to be called in for advice and suggestions by the county agricultural agents and farm owners. As the farm woods is, in their minds, only a minor part of the farm, so the forester or the wildlife expert is only a minor technician, quite unfit for large responsibility in land management and policy determination. One can imagine the attitude of this group toward the many important problems of forestry and wildlife management confronting the TVA.

3. That the TVA is a unique effort on the part of the people of the United States, through the federal government, to solve the major problems of a great watershed. The advocates of this third concept propose to approach the job with a scientific attitude. They have been assembled from all over the U. S.; they are spending most of the money—and are doing most of the work. To them the TVA is simply a large federal organization, with a very clearcut, definite job. They believe that the other federal and state organizations are important, and that the TVA should coöperate with them fully, but that TVA should not turn over to them, however, the work of the TVA itself.[30]

An examination of the record indicates that the TVA forestry division has not had an exclusive orientation toward any given set of local institutions. Thus, while lacking confidence in the extension service as a whole as far as forestry work is concerned, the division has made efforts to strengthen the hand of the extension forester within his own organization. There has also been coöperation with the vocational agriculture teachers, and good relations with the Forest Service. The dominant note seems to be one of promoting the professional interests of forestry, wherever this may lead. To some extent, this has in practice involved an effort to build up the state conservation departments and forestry divisions, but on the whole these have not been sufficiently strong to form a firm constituency. Moreover, the TVA foresters would tend to avoid being tied too closely to the conservation departments, for that might result in exclusion from work on farm woodlands. In general, it may be said that the TVA forestry division has followed perspective 3 listed above in the quotation from Richards.

With the victory of the Lilienthal–H. A. Morgan bloc in the TVA as a whole, and the concomitant departure of Richards, pressure for the reorientation of the forestry division along lines desired by the agriculturists was renewed. Despite the change in leadership, the forestry division retained its direct-action viewpoint, though in a somewhat chastened form. In the general reorganization of the Authority, the Forestry Relations Department was brought under the supervision of a Chief Conservation Engineer, whose office functioned as a convenient {150} mechanism for the exercise of control over the foresters to implement this reorientation. In 1937, on the initiative of the TVA agriculturists, the colleges were requested by the Authority to recommend a forestry policy. Not unnaturally, the institutions recommended that TVA follow the same procedures in the farm forestry program as were utilized in the agricultural work. It is significant for understanding the meaning of such formal mechanisms that the Agricultural Relations Department was able to use (1) this formal recommendation from the colleges, (2) the original three-way memorandum of understanding, and (3) the precedents established in its own activities, as arguments to buttress the attempt to bring the TVA foresters into line with its own procedures.

The work of the Forestry Relations Department with CCC camps was discontinued in 1942. This, in addition to other wartime developments, provided the occasion for the agriculturists to renew the attempt to alter the orientation of the foresters. After conferences with representatives of the Forestry Relations and Agricultural Relations Departments, the Chief Conservation Engineer "laid down the line," insisting that the foresters carry out the obligations of the Authority so as to strengthen the land-grant colleges "in the performance of their responsibilities to the people of their states and so as to avoid duplicating and confusing field administration." Specifically, he said that this was to be done by adding additional personnel to the extension service staffs. It is safe to say that the foresters were less than enthusiastic about this injunction, and by 1943, a total of only about five men were either assigned to the administrative jurisdiction of the extension services or hired by the latter with salaries reimbursed by TVA. Moreover, the contract with the University of Tennessee for the employment of two foresters (assistant county agents) indicated that the Forestry Relations Department was reserving more control over the administration of its contracts than was customary in the Agricultural Relations Department. This contract included provisions for TVA approval of the assistant agents selected and for inspection by TVA of their technical performance.

In general, even after the departure of Richards and others of the A. E. Morgan group, the TVA foresters remained unconvinced of the value of the procedure advocated by the agriculturists. This is a matter of method in relation not only to the extension services, but to all local agencies. Thus the foresters, in supporting and strengthening the state conservation departments, do not favor grants-in-aid or subsidies by TVA. They follow rather the way of building public support for appropriations to these agencies on a county and state level, as well as for

{151} further funds from the federal government under the provisions of the Clarke-McNary Act. Favorable public sentiment is encouraged through a forest-fire educational program carried on at the county level, and by official TVA representations to the President as to the states' need for increased funds for work in forestry. This general orientation, added to the initial lack of faith in the ability of extension service to carry on a forestry program, makes them loathe to turn over personnel to that organization. But because of the prestige of the grass-roots method within the Authority and the power of the agriculturists, the TVA foresters seem to be fighting a losing battle.

2. In exercising its function as the representative of a constituency, it is convenient for the TVA agriculturists to make themselves the primary or even exclusive channel of the Authority to the colleges, which insures that direct contact will be made through individuals who will view the joint activities with friendly eyes and frame their reports in terms which will uphold the established relationships. Two examples may suffice. In 1941, the TVA Director of Agricultural Relations attempted to restrain representatives of the TVA Comptroller from examining field reports of the extension service at the institutions, in connection with audits of reimbursements under coöperative contracts; he insisted upon a literal interpretation of a clause in the project agreement which states that "during the progress of the work the Chief of the Agricultural Division of the TVA, or his representative, shall have access to any records or reports pertaining to the progress of the work hereunder." Thus the agreement so framed is a prop of the constituency relation, protecting the colleges against possible unfriendly attacks. Both the colleges and the TVA agriculturists interpret their agreement as not with the TVA as a whole, but with the Department of Agricultural Relations. This tactic may also be useful to the TVA agriculturists in holding the institutions in line, lest members of the college staffs be overly coöperative with other branches of TVA and thus serve to loosen the established ties. The restriction of contact serves not only to protect the colleges but also to preserve the existing constituency relation.

To this more or less routine effort we may add as a further example the attempt of the TVA agriculturists to formally establish themselves as the coördinators of all activities by the Authority which affect agriculture in any way. This is the proposition, mentioned above (p. 116), for the promulgation by central management of an Administrative Code which would solidify the position of the land-grant colleges and give the Agricultural Relations Department authority to approve or disapprove {152} all proposals initiated by other departments in the TVA and affecting agriculture or rural people. With the Authority operating in a predominantly rural area, the significance of such control exercised by a group in TVA with special commitments to outside elements would be very great. Up to June, 1943, no action had been taken by central management on this proposal. If adopted, the code would give the Agricultural Relations Department a measure of control over the activities of a number of TVA departments whose activities affect rural people, including Commerce (agricultural engineering developments), Forestry Relations, Regional Studies, Power Utilization (rural electrification), Health and Safety (malaria control work in rural areas), and Reservoir Property Management (contact with farmers in the administration of TVA-owned lands).

3. The TVA agriculturists have, during the history of the Authority, found a number of opportunities to defend the extension service organization and outlook from encroachments by other departments in the Authority. This was more evident in the early days of the TVA, when efforts to think in terms of regional planning in its broader sense were more common. The agriculturists were known as the bitter enemies of the Land Planning Division, composed of what they called, pejoratively, "the geographers." Reports issued by other departments affecting rural areas were subjected to severe criticism by the agriculturists, and in one case at least, as late as 1939, an attempt was made to ban the distribution of a study of rural zoning which had been carried out not by the TVA alone but coöperatively with the Bureau of Agricultural Economics of the U. S. Department of Agriculture, the Tennessee State Planning Commission, and the Hamilton County (Tennessee) Regional Planning Commission. The TVA's Departments of Regional Planning Studies and Forestry Relations of TVA had participated. In the same year, another report compiled by the Regional Studies Department on the probable effects of one of the TVA projects upon economic and social life in the area was criticized by the Director of Agricultural Relations:

1. There is little, if any recognition of the prior right and responsibility of people in local areas to take part in the planning and execution of the project under consideration.

2. The report does not capitalize on the judgment and experience which resides only in the local areas and can be made available only through organized study and analysis by the local people.

3. The conduct of investigations into the private affairs of individuals without their collective concurrence goes beyond what is proper for a Federal agency and abridges and undermines the relationships which have been accepted as policy by the local, state, and Federal agencies. {153}

4. Without the collective judgment of responsible organized groups, it is doubtful if information obtained is reliable and trustworthy as a basis of action of such far-reaching consequences in local areas.

I reiterate my attitude, many times expressed on similar matters, that studies made and conclusions drawn with reference to agricultural situations and problems in local areas should be done by, and largely for, the people themselves, acting through their local, state, and Federal educational agencies.

The implication of these remarks clearly seems to be that only such institutions as the land-grant colleges may rightfully prepare reports on matters affecting farm people. It is doubtful that the Director of Agricultural Relations would consider a local planning commission, organized with the aid of some other TVA department, a bona fide "local institution."

4. Examples of the furtherance of specific policies of the land-grant colleges by the TVA agriculturists will be detailed in chapters v and vi below, indicating that the role and character of the TVA as a whole have been shaped, at least in part, by the coöptative relationship in the field of agriculture.

In conclusion, we may consider the mutual dependence of the Agricultural Relations Department as agent, and the land-grant colleges as constituency. Two views are presented by persons close to the program. One suggests that the colleges "don't really give a damn" about TVA or even the money which it turns over to them—that the colleges actually have a whip hand over the Agricultural Relations Department in the possibility of their withdrawal from the joint program. Since the department is so deeply committed to the theory that it is impossible to carry on a program apart from the colleges, such an eventuality would be disastrous. On the other hand, it has been suggested by a leading participant that Director McAmis "has strong control over the state extension services, who fear a possible withdrawal of TVA funds." Doubtless both the constituency and its agent are mutually dependent upon each other, but it seems probable that the dependence of the TVA agriculturists on the preservation of the established relation is the greater. A reorganization of TVA agricultural activities might change the form of coöperation with the extension services—and vitiate some of the consequences of the earlier relationship,—but coöperation in some form would continue. A change in policy by the top TVA leadership might result in the removal of the existing group in control of the agricultural program—though with some sacrifice of TVA's position in the Valley,—but it could not seriously affect the administrative existence of the extension service.

----

[1] See especially Gladys Baker, *The County Agent* (Chicago: University of Chicago Press, 1939); also Russell Lord, *The Agrarian Revival: A Study of Agricultural Extension* (New York: Association for Adult Education, 1939); V. O. Key, *Administration of Federal Grants to the States* (Chicago: Public Administration Service, 1937); G. A. Works and Barton Morgan, *The Land-Grant Colleges*, Staff Study No. 10, Advisory Committee on Education (Washington: U. S. Government Printing Office, 1939).

[2] Baker, *op. cit.*, pp. 60, 102 ff.

[3] Lord, *op. cit.*, p. 110.

[4] See below, p. 140.

[5] Baker, *op. cit.*, pp. 15 ff.

[6] The number of states having such a formal tie has been steadily declining. The formal connection, however, is not necessarily the significant point. In some states where the legal tie exists, it may not be meaningful. However, where the legal link does not exist, the informal bonds may continue strong. In 1947 a spokesman of the National Farmers Union claimed that "in more than half of the States of this Nation, the Extension Service is not a free, educational agency, but so closely connected with a single organization within agriculture that the two are sometimes indistinguishable, one from the other." See *Hearings* on S. 1251, Senate Committee on Agriculture and Forestry, 80th Cong., 1st sess. (1947), p. 385. Yet legally the tie is to a "farm bureau" without reference to the American Farm Bureau Federation, so that theoretically a local group need not affiliate with the AFBF. The Ohio Farm Bureau Federation has often been opposed to the policies of the national organization, and even the New York Farm Bureau Federation opposed S. 1251 (National Soil Fertility Bill) though the bill was strongly supported by the national headquarters. Nevertheless, the AFBF is one of the great lobbies of the United States, and its intimate relation to the agricultural extension service is one of the striking characteristics of the structure of American agricultural organization.

[7] Baker, *op. cit.*, p. 141.

[8] This agreement is reproduced as an appendix to Works and Morgan, *op. cit.*

[9] Memorandum No. 893, Office of the Secretary, U. S. Department of Agriculture, Washington, D.C., March 21, 1941 (mim.).

[10] See Baker, *op. cit.*, pp. 212-213. Presumably as a response to criticism suggested by statements of this sort, a number of studies have attempted to evaluate the extent to which the extension service reaches different strata of the population. M. C. Wilson ("How and to What Extent is the Extension Service Reaching Low-Income Farm Families?" U. S. Department of Agriculture Extension Service Circular 375, December, 1941) concludes that on the county level "slightly more of the farm families in the average and above-average socio-economic segments than of relatively disadvantaged farm families in the same areas are being reached by Extension. The differences, however, are not great. In fact they would seem to be even smaller than might be naturally anticipated in a situation involving voluntary participation of people in Extension teaching activities. Extension must render reasonable service to those progressive people who already have the desire for information. This frequently limits the attention a limited county staff can devote to arousing a desire for information on the part of those who do not yet have that desire." The term "reached" used here is very important, and it is used in Wilson's report to mean "in some

measurable way." However, D. L. Gibson ("The Clientele of the Agricultural Extension Service," Michigan Agricultural Experiment Station *Quarterly Bulletin*, Vol. 26, May, 1944) analyzes the problem without leaving the term "reached" ambiguous. Gibson finds that in the Michigan counties studied the differences among socioeconomic groups were not great in respect to such passive participation as receiving mimeographed material, but the differences were considerable when more active types of participation, such as personal visits and running test plots, were considered. Gibson concludes that, for the area studied, "the so-called 'disadvantaged class' of farmers, those of lower socioeconomic status, participated decidedly less in the agricultural extension program than did those of high status." As Gibson suggests, this is no reflection upon the efforts of county agricultural agents. It is important objectively, however, in considering the character of the extension service. See also C. B. Hoffer, "Selected Social Factors Affecting Participation of Farmers in Agricultural Extension Work," Michigan State College Agricultural Experiment Station Special Bulletin 331 (June, 1944); and W. A. Anderson, "Farm Youth in the 4-H Club," Department of Rural Sociology, Mimeographed Bulletin No. 14 (Ithaca, New York: Cornell University Agricultural Experiment Station, May, 1944).

[11] Recently renamed, National Farm Labor Union, affiliated with the American Federation of Labor.

[12] From a sample contract.

[13] Reimbursements are made for detailed items. The possibility of making lump-sum grants to the colleges for executing a delegated responsibility has been explored, but rejected. The states need not match funds provided by the Authority, though such a procedure was included in the original Smith-Lever Act of 1914. See n. 16, below.

[14] The pattern developed between the Department of Agriculture in Washington and the land-grant colleges in administering the Smith-Lever Act of 1914 under which agricultural extension work was inaugurated on a national scale.

[15] It will probably bear reemphasizing that this study is not concerned with appraising the grass-roots method in terms of its formal objectives; interest is restricted to the organizational implications which throw light on the interaction of the Authority and the colleges.

[16] This does not mean, of course, that stimulation has not occurred. On the contrary, there is evidence that in some fields the colleges have expanded on their own from the core represented by the TVA-subsidized program. It is not known to what extent some other method, including lump-sum grants and matching provisions, would have affected the colleges' initiative. The point here is simply that the objective meaning of the doctrine of stimulation sustains a loose administrative relationship. It is noteworthy that reimbursement for detailed items, while superficially representing a measure of control by the Authority, is in fact more congenial to those who

advocate a loose relationship than is the lump-sum grant. The use of the latter procedure immediately raises the question of a system of fiscal control and work planning, as well as the possibility of instituting a matching requirement.

[17] P. 129 ff.

[18] See above, pp. 99 ff., for data on extent of the program.

[19] See table 4, p. 102.

[20] A more detailed discussion of the county soil associations will be found below, pp. 230 ff.

[21] See table 7 above, p. 104.

[22] Pp. 117-124, above.

[23] *Hearings* on S. 1251, Senate Committee on Agriculture and Forestry, 80th Cong., 1st sess. (1947), p. 297.

[24] "Progress and Possibilities for Further Progress on 50 Unit Test-Demonstration Farms in the Valley of East Tennessee, A Preliminary Report," mimeographed report issued by the University of Tennessee (June, 1942).

[25] The decline in effectiveness of the test-demonstration program in high tenancy areas was attested to by a district agent of the Tennessee Extension Service.

[26] See pp. 119 ff. above.

[27] *Hearings*, Joint Committee to Investigate the Adequacy and Use of Phosphate Resources of the United States, 75th Cong., 3d sess. (1938), p. 202.

[28] *Hearings* on Disposition of Muscle Shoals, Senate Committee on Agriculture and Forestry, 70th Cong., 1st sess. (1928), p. 79.

[29] See Avery Leiserson, *Administrative Regulation* (Chicago: University of Chicago Press, 1942), p. 9; also V. O. Key, Jr., "Politics and Administration," in L. D. White (ed.), *The Future of Government in the United States* (Chicago: University of Chicago Press, 1942), p. 151: "In one respect individual parts of the administration tend to become like private pressure groups in that they have their own particularistic and parochial interests to defend and promote.... The association of the agency with its clientele sometimes makes for a 'representative bureaucracy.' The policy drives of the hierarchy arising from the immediate interests of its members are reinforced and colored by the power and wishes of outside groups concerned with the work of the agency. Through administrative determination of delegated policy questions and through administrative influence on new policy, the desires of private groups may be effectively projected into the governmental machinery. In extreme circumstances something in the nature of a guild may be approached."

[30] E. C. M. Richards, "The Future of TVA Forestry," *Journal of Forestry*, 36 (July, 1938), 644-645.

CHAPTER V

THE STRUGGLE IN AGRICULTURE AND THE ROLE OF TVA

I'm not opposed to security spelled with a small 's'; but when you use a large 's,' it leaves me cold.

DIRECTOR OF AGRICULTURAL RELATIONS, TVA (1943)

THE EXERCISE of administrative discretion gives life and meaning to the abstractions of policy and doctrine.[1] The injunctions of a statute, the directives of a central management mobilize action on the assumption that common organizational goals exist and that these goals are effective in shaping rational administrative behavior. But this assumption breaks down as, in the exercise of discretion, officials are faced with the need to deal selectively with their environments. For as they come upon an existing social situation, individual administrators find it difficult to restrict their interest and involvement to the formal goals of the policy they are executing; they tend to be involved wholly, bringing to bear their own fractional interests upon day-to-day decisions. These special commitments may be well organized and consistent, or diffuse and sporadic. They range from the extreme case, in which the whole allegiance of the official is to some outside group or interest, to the relatively mild and thoroughly normal attention which a field representative may give to bolstering his position within the organization. The former case may require drastic correctional measures, whereas the latter may involve only minor changes in the mechanics of central control. But whatever the forms, or the intensity of their expression, the centrifugal tendencies inherent in the delegation of discretionary powers must be recognized and taken into account by those responsible for the integrity of the whole organization.

In the analysis of this centrifugal tendency, we are led on theoretical grounds to the following considerations: (1) The organization may enter upon a preëxisting social situation in which there are loci of power which will not permit themselves to be ignored as activities in which they have an interest are initiated. (2) Intervention may occur in a context of controversy among forces within the area of operation, so that the problem of influencing the controversy may be posed. (3) The intervening organization may be actually or potentially split into factions, {156} thus enhancing the possibility that one group within the organization may exercise whatever discretion may be delegated to it in terms of tangential goals and commitments. (4) As the discretion of a part of the administrative apparatus is exercised in the name of the organization as a whole, the latter may find itself committed to a policy or a set of relationships which it neither anticipated nor desired. Thus the parochial interests of a subordinate part of an organization must be viewed in the light of the social situation within which operations are carried on and in terms of long-run consequences for the position of the organization as a whole.

If we combine these theoretical considerations with the preceding discussion of the TVA's agricultural machinery, it will be possible to draw out more fully the meaning of the grass-roots idea as it has worked out in practice. This meaning is explicated as consequences are traced and the practical force of the doctrine observed. The consequences are objective, in the sense that they have been determined by forces generated willy-nilly from the initial relationships established; they are unintended in the sense that they are not the predicted products of executive action. The grass-roots policy suggests that the program of a regional agency may be channeled through existing agencies, but does not, on the level of doctrine, take into account the possible tangential results of such a procedure.

There are, as we have noted, three levels which define the "decentralized administration of federal functions." These are (1) the location of an agency in the area of operation, with sufficient managerial autonomy to enable it to deal effectively with local conditions in local terms; (2) the method of working with and through local agencies in the execution of a program; (3) the encouragement of participation by the people themselves in administration, primarily through the use of voluntary associations. The first of these conditions raises the problem most directly of the relation of such an agency as TVA to the rest of the federal structure. The second condition seems to be independent of the first, concerned as it is with the relation of the agency to the local institutions. But this mutual exclusion is only apparent. As selection and intervention with respect to the local area occur, it becomes clear that many local decisions will be made in terms of their national consequences. Local intervention may offer an opportunity to influence a far-reaching decision which at first sight is irrelevant to the specific discretionary problem posed.[2] Consequently, it may be anticipated that the concrete {157} exercise of discretion, in relation to the principle of working with and through existing institutions in the area, may seriously affect the relation of the agency as a whole to the federal system. This inference is borne out by the TVA experience, as will be made clear in the course of this chapter.

INTERAGENCY RIVALRIES IN AGRICULTURE

The initial organization of the TVA in 1933 was immediately faced with the existence of the land-grant college system in the field of agriculture. As we have seen,[3] the potential issue was resolved in advance by the absorption of the TVA agricultural program within that of the extension service and, more specifically, by the establishment of the constituency relationship between the colleges and the Agricultural Relations Department of TVA. But the Authority entered upon a social situation which not only contained preëxisting organizations having established prerogatives but was, in common with the rest of the nation, the scene of a protracted controversy among contending social forces for control over the agricultural machinery of the United States government. In this controversy, the TVA as a whole, borne along by the constituency relation already established, became committed to one of the contending sides.

For almost a generation after the establishment of the agricultural extension service in 1914, the contact of agencies of the government with the farm population remained relatively simple. The county agent system represented the main channel through which information and, from time to time, exhortation was brought to the farmer.[4] Twenty years of operation, added to the earlier years of preparation, served to stamp the extension service organization and personnel with a traditional character, inevitably shaping its conception of the problems with which it might be faced.[5] The stress on education made for rather easygoing operations, aiming at a gradual improvement of farming methods rather than any immediate ameliorative measures. Knowledge, rather than social or economic organization, was considered the limiting factor: the objective of "growing two blades of grass where one grew before" was uppermost. The spread of knowledge by example and demonstration {158} smoothed the way to the substantial farmer of the county who could readily participate in new ventures and gradually set the pace for his neighbors. In this way, the county agent system developed a set of predispositions and habits of work which clashed with the demands of those who later called for aggressive action. Moreover, the association of the agents with the more prosperous farmers tended to compromise their position as class forces within agriculture began to erupt.[6]

The break in prices and land values which followed the heady expansion during World War I resulted in a persistent agitation for government aid to the farmers. This pressure found its greatest response after the advent of the New Deal administration, which opened a new era in agricultural policy. A combination of measures for relief, production control, soil conservation, and vastly expanded credit facilities served to pose a series of new and pressing problems for those involved in the network of administrative and political organizations in agriculture. These new problems crystallized around two basic issues: (1) the emergence of internal conflict among the farmers, according to distinctions of economic class, and reflected in the alignments of farm organizations; (2) the change in the character of federal intervention in agriculture, reflected in a challenge to the older governmental institutions, as new agencies came upon the scene. These two issues were merged in the actual struggle, in which agencies functioned as weapons in class conflict.

The American Farm Bureau Federation was a leading contender in the agricultural struggle during the decade following 1933, and the issues in which this organization was involved most clearly defined the lines of controversy. The special role of the Farm Bureau was determined by its identification with the well-to-do and "industrial" farmers as a class, and with the extension service system as its organizational weapon. The class attitude was disclosed most clearly in the opposition of the Farm Bureau to the Farm Security Administration. The latter directed its program toward the underprivileged groups in agriculture, especially the tenant farmers, and was supported by the Farmers Educational and Co-operative Union of America as well as by such smaller groups as the Southern Tenant Farmers Union and the Farm Holiday Association. The Farm Bureau bitterly attacked the entire program of the FSA and demanded that its functions be either scrapped or turned over to other agencies.[7] Some light on the class basis of this attack {159} is cast by the following statement attributed to a high official of the FSA:

> Two fights over income are going on in American Agriculture. One has to do with Agriculture's getting its fair share of the total income of the nation—the "parity" idea that the Farm Bureau has championed for years. The AAA is the key agency in this, which explains why it has been so long and bitterly fought by many persons *outside* the ranks of Agriculture.
>
> The other fight is *inside* Agriculture's own family—over distribution within the family of whatever income Agriculture gets. FSA figures sharply in this, as champion of the group that has long been on the small end. It follows logically that the bitterest enemy of FSA in this fight would not be a disinterested group outside Agriculture's fold, but the established, well-to-do group within the family that has been enjoying the biggest share in the past—and the Farm Bureau Federation tends to represent that group.[8]

In more specific terms, another supporter of the FSA stated:

> The real reason for the bitterness lies in the fact that certain types of agriculture—the cotton plantation, the truck farm, the orchard—require a large and easily available supply of cheap labor. This labor is drawn mostly from the poorest, the wretchedest, and the most shiftless of all our farm people. These are also the people whom the FSA undertook to "rehabilitate." The oftener the FSA succeeds—and they have been at it for nearly ten years—the more the commercial farmer's labor supply diminishes. And, to the extent that the "rehabilitated" become more than subsistence farmers, the more competition there is for the market.[9]

The latter interpretation is supported by a reading of testimony at congressional hearings on the FSA. Thus, in March, 1943, the National Cotton Council, of which the American Farm Bureau Federation is an important constituent, attacked the FSA on the grounds that

> . . . this agency is so functioning and so conducting its activities as to promote gross inefficiency in the matter of culture and production of cotton and cottonseed; to seriously impede cost of production . . . to lower the morale of farm workers . . . to threaten, disturb, and disrupt economic and social conditions and relationships throughout the Cotton Belt; to promote class distinctions, hatred, prejudice and distrust within the Cotton Belt; to threaten those who produce cotton and cottonseed on a commercial basis; to depress morale of cotton farmers throughout the belt, and ultimately to destroy the business of farming as a free enterprise and a respectable means of earning social and economic security by American farmers.[10] {160}

On another occasion, the Farm Bureau representative introduced as evidence against the FSA a letter from a landowner protesting that the FSA was "taking away tenants from landowners."[11] Again, President Edward A. O'Neal of the Farm Bureau registered his support of protests against the FSA practice of including poll tax obligations among allowable expenditures of tenant farmers in Alabama.[12] These considerations, among others, indicate that opposition to the FSA has been programmatic and substantive rather than merely organizational or administrative.[13]

The organizational phase of the controversy has, however, been equally significant. As a mass organization, the Farm Bureau has been highly conscious of the need to protect its avenue of access to the farm population. To the extent that the Farm Bureau is identified with whatever benefits farmers receive from governmental programs, its chances for a large membership, and for control over farmer sentiment, are vastly enhanced. This opportunity was for many years upheld by the special relation of the Farm Bureau to the extension service organizations.[14] The president of the Iowa Farm Bureau Federation is reported to have stated that "without the extension service the Farm Bureau would die."[15] The Farm Bureau is the consistent champion of the extension service in the protection and enlargement of its appropriations and in defense of it against competing agencies. The county agent system functions to build and preserve the Farm Bureau membership. In defending the extension service, the Farm Bureau fulfills an elementary condition of its own survival.

The Farm Bureau's defense of the extension service may, however, assume proportions which go beyond the needs of the college organizations themselves. The land-grant colleges are well established, and in many areas need not fear the existence of other agricultural agencies {161} so long as their own integrity as educational institutions is preserved. Other agencies, indeed, may even overshadow the colleges in importance without producing panic among the latter, since the colleges continue to have an enduring function. But for the Farm Bureau the extension service represents its avenue of access to the farm population and, to the extent that that access is compromised or dwindles in importance, the Farm Bureau leaders may expect that farmers will cease to turn to them for leadership. The Farm Bureau is therefore driven to push the extension service into accepting ever increasing responsibilities. The problems posed as governmental agencies come to function (or threaten to function) as quasi-political machines have involved the Farm Bureau and the land-grant colleges in a series of persistent difficulties.

The Farm Bureau's role as a representative of agricultural interests in Washington has moved it to sponsor new modes of intervention by the federal government for the benefit of the farmers. Perhaps the farthest reaching of these in recent years has been the Agricultural Adjustment Administration. The Farm Bureau "fathered" the AAA Act of 1933[16] and was one of the chief supporters of the New Deal agricultural program in its early and major phases. But its dilemma consisted in this: in promoting the new program, it could not avoid the construction of a new organization, the AAA. The latter, growing strong, tended to have influence upon the farmers in its own right, and thus to vitiate the prestige and strength of the Farm Bureau. This problem was foreseen, for since the early days the Farm Bureau insisted that administration of the program be channeled through the extension service. Much was done toward that end,[17] but the large amount of patently non-educational work involved and a growing disaffection in Washington with the loose Smith-Lever relation of the Department of Agriculture to the states[18] resulted in the construction of an independent AAA organization. This included a network of county and community committeemen participating in the administration of the program. These committeemen numbered approximately 90,000 and were paid about $7,500,000 out of AAA funds in 1941.[19] As the years went by, the attitude of the Farm Bureau toward the AAA cooled markedly, and by 1943 the {162} undercurrent of antagonism broke out in open conflict.[20] On April 1, 1943, the *Tennessee Farm Bureau News* carried the following headlines:

> TRIPLE A LEADERS GUNNING
> FOR AMERICAN FARM BUREAU
> MOUNTING EVIDENCE REVEALS
>
> OUT TO DESTROY
> ORGANIZATION
> O'NEAL SAYS

It was charged that AAA officials were attempting to antagonize farmers toward the Farm Bureau because of the latter's stand against the continuation of AAA benefit payments. While the occasion for the outburst[21] was the stand of the Farm Bureau on this particular issue, the fundamental problem was that of organizational rivalry.

The Farm Bureau's answer to its dilemma with respect to AAA was to insist that all educational, informational, and advisory services needed in connection with the AAA program in the states, counties, and communities be turned over to the extension service. This would insure that the agency which dealt with the farmers would be friendly to the Farm Bureau and help it to retain its mass support. This solution was proposed consistently for all the new agricultural agencies which threatened to construct an organization reaching down to the farm population and which the Farm Bureau could not expect to control. Thus the Farm Security Administration represented not only a class antagonist but an organizational rival. The FSA did not carry on its work through the extension services; it built its own field organization, with FSA representatives at the county level dealing directly with farmers. This presented the dangerous possibility that the county agent's office might cease to be the primary means of access to the farmer—a possibility inherently inimical to the interests of the Farm Bureau. The Bureau consequently demanded that FSA be dismembered, with its farm and home management services transferred to the extension service. {163}

The Farm Bureau also saw a threat to its position in the growth of the Soil Conservation Service, another organization having its own men in the field with direct access to the farmers. The program of the Farm Bureau with respect to SCS, though lacking the class antagonism which marked its attitude toward FSA, still demanded that the land-grant colleges be utilized, rather than the SCS. Under its program,

> . . . the work of the Soil Conservation Service in Washington would he handled by the appropriate agencies which were established within the Department of Agriculture to do such work; regional offices would be abolished; research and experimental work in soil conservation would be transferred to the State Agricultural Experiment Stations and carried on in coöperation with the U. S. Department of Agriculture; educational, demonstrational, and advisory assistance to farmers and Soil Conservation Districts would be transferred to the State Extension Services and carried on in coöperation with the Department of Agriculture.[22]

In this way, the Farm Bureau consistently defended its most vital asset —its nearly exclusive access to the farm population through the machinery of the agricultural extension service.

The dilemma of the land-grant colleges had the same basis as that which plagued the Farm Bureau, but the alternatives were different. The challenge of the new agencies seemed to demand that the colleges offer their own organizations for the administration of the action programs; and this impulse was heightened by the desire of the Farm Bureau that they do so. At the same time, the colleges desired to maintain their integrity as educational institutions, leaving the task of administering a planned agricultural program to others. Many of the colleges desired to avoid being identified with an agricultural program which might be too closely linked to the current administration in Washington, or which might force the colleges to be the instruments through which penalties against their constituent farm populations might be enforced. Nor did the colleges respond uniformly. In the Southern states, the weakness and relative poverty of the institutions made the new program especially welcome. But in some of the more wealthy farm states, as in Iowa,[23] there was a stronger tendency to eschew administrative responsibility for the action programs.

Despite a certain ambivalence in the attitude of the colleges, it was clear that on the whole they had little sympathy for those newer trends in the Department of Agriculture which reflected a basic change in {164} attitude toward the methods by which agricultural programs were to be administered. Many of the newer officials in the Department had little confidence in the extension service[24] and presumably the more extreme elements served to create a favorable climate of opinion for a revamping of the official procedure. Milton S. Eisenhower said of this:

> Most of the programs involved in this new comprehensive policy required planning and action by farm people. They required also coöperative action by Federal, State, and local agencies. But the coöperative action was of a different kind. It departed from the traditional relationship of the grant-in-aid type that had existed between the Colleges and the Department in research and extension work. In framing the new laws the Congress, seeking to bring about simultaneous nation-wide action, placed on the Secretary of Agriculture the responsibility for their administration; the new laws did not authorize him to transfer his responsibility to any other person or agency. Naturally, the laws authorized coöperation.[25]

Thus it was not only the problem of facing competition in the field, but also that of a change in the traditional and secure policy of the department in the administration of its functions, which disturbed the land-grant college chiefs. This issue created an atmosphere of controversy in the field as well as what has been referred to as a split within the Department of Agriculture itself in Washington.[26] To be sure, the more dramatic aspects of this controversy within the Department may lead to overfacile conclusions as to basic divisions in function. The "new" Department is a product of the "old," and the action programs are the result of a long evolution within the Department.[27] However, the spirit of controversy, though generated by the relatively passing and temporary needs of organizational prestige and survival, is a significant factor which must be taken into account in analyzing the behavior of interested individuals and groups as they exercise administrative discretion.

TVA AND THE FARM SECURITY ADMINISTRATION

The leadership of the TVA's agricultural program has been an institutional leadership. As such, it has conceived of its own role as larger than the administration of a fertilizer program, justifying a self-conscious {165} approach to the evolution of the pattern of organizational control in agriculture. Keenly aware of its function as representative within the TVA of the land-grant college system, the Agricultural Relations Department has scrutinized all coöperative arrangements affecting farm organizations and rural areas with an eye to the possible consequences for its constituency.[28] In defending the colleges within the Tennessee Valley, moreover, the TVA agriculturists have found it necessary to take the side of the system of land-grant colleges as a whole, in national terms. This necessity, coupled with the personal history of the participants and the working relationship with the national leadership of the Farm Bureau,[29] has made them acutely aware of the over-all controversies which have been outlined in the preceding pages. Consequently, in exercising delegated discretionary powers, and in taking advantage of opportunities within the Authority for influencing policy on tangential matters, the agriculturists have sought to commit the Authority as a whole to the cause espoused by the Farm Bureau and the extension service. This has been done not in speeches or testimony or other forms of public witness, but through the "dry" method by which the weight of an organization is wielded: the establishment of precedents, the acceptance or rejection of opportunities for joint effort, the subtle intimations of disaffection put on paper, and the more plain verbal injunctions which permit field agents to know the attitudes and desires of headquarters.

The handling of proposals for coöperative relations among agencies is one of the touchstones by which real attitudes may be determined, and is a field within which discretion normally has wide play. "Coöperation" is a term which remains unanalyzed so long as its use in administrative parlance neglects to specify references, procedure, and consequences. A long list of agencies with which coöperative relationships are maintained may be as meaningless as it is casually impressive so long as it represents an amalgam of the simple with the complex and slurs over the real and significant distinctions among the relationships established. This substitution of shadow for substance is a convenient means of fulfilling the formal demands of a statute or a Congressional committee, as well as the need for a shield against criticism of some specific relationship. {166}

Insofar as coöperation bespeaks more than words—though these too are important—it has meaning for the future of the organizations involved. Especially when negotiation is infused with motives which go beyond the substantive programs in question, that future is at stake. If an official is hostile to the organization with which coöperation is proposed, he will consider the possibility that the joint program will serve to strengthen that organization through: (1) the precedents to be established, which may be difficult to overcome at a later date; (2) the possible influence of that organization upon his own rank and file; (3) the extent to which funds allocated to him may leave his control in the course of the joint activity; (4) the danger that public credit may be shared, or that the two organizations will become confused in the public mind; (5) the possibility that access to client publics, not previously available, will now be afforded to the coöperating organizations; and (6) the establishment of an actual machinery (personnel, facilities, reciprocal commitments) which might become intrenched.[30] If the attitude is friendly rather than hostile, and especially if there is interest in actually building the coöperating organization, then precisely these considerations will strengthen the case for coöperation. But if the official is organizationally self-conscious, he will not ignore them.

It is not the least of the paradoxes characterizing the TVA that there should have been so little in the way of rapprochement with the Farm Security Administration. Both agencies seemed to be part of the more experimental phase of the New Deal, and both derived support from the liberal-labor movement in the United States. It was perhaps not unreasonable to expect, therefore, that the two agencies would have a very close relationship, each contributing to the other's strength and prestige. But this is to reckon without the split in the character of TVA: carrying out its mission in the field of electric power, but permitting the agricultural program to be controlled by representatives of the existing institutions of the area. In fact, as we have seen, the actual alliance of the TVA agriculturists was with the bitterest enemy of FSA—the Farm Bureau. Consequently, the Authority as a whole was placed in the anomalous position of rebuffing what many would have considered to be one of its natural allies.

The policy of rebuffing proposals for coöperation with FSA stemmed from the Agricultural Relations Department of TVA. Other divisions of the Authority were willing, even eager, to work with the FSA. But {167} constant pressure from the agriculturists, whose control over the major subject-matter department interested in rural people placed them in a strategic position, led to a virtually complete elimination of FSA from joint activity with the Authority. This objective was accomplished in rather subtle ways, so that no complete defense or explanation has been spread upon the record. It was made easier since no formal relationships on a general scale had ever been initiated, for the extension service group had taken control over the agricultural program from its inception. Consequently, no open break was necessary, nor any extended discussion within the Authority. Nevertheless, it is possible to set forth some details on the methods used by the agriculturists to swing the TVA away from coöperation with the Farm Security Administration.

1. The attitudes of leading personnel within the Agricultural Relations Department are obliquely hostile to the Farm Security Adminis-

tration and other federal agencies in agriculture which do not follow the grass-roots method. This is expressed in general terms, as in talks to personnel stressing the dangers of increased centralization and the threat to local institutions from new ventures by the federal government; outspoken support of the Farm Bureau position—during days of national controversy—that the county agent should be placed in charge of all agricultural programs at the county level; and more rarely, relatively explicit references such as that of the Director of the Department that he was "scared of 'security' spelled with a large 's.'" Considering also the attitudes outlined above,[31] it is hardly surprising that the TVA agriculturists would be cold to proposals for coöperation with FSA.

2. Officials of the TVA Forestry Relations Department were eager to work out a joint program with the Farm Security Administration, partly because of the resistance felt by foresters on the part of the extension service. Some contacts were initiated in 1938. By that time, however, the TVA foresters had already been bracketed with the Agricultural Relations Department under the supervision of the Chief Conservation Engineer. In November, 1938, a discussion was held including representatives of the foresters, the Director of Agricultural Relations, and the Chief Conservation Engineer, at which the question was posed of whether TVA "should recognize Farm Security." According to one source, Director of Agricultural Relations McAmis said "No," to which the Chief Conservation Engineer added, "Whatever McAmis said on the question must go." More recently, an attempt was made to route assistance to FSA through the State Forester in Tennessee, and some {168} work was done on FSA farms in this way, but a TVA forestry official was told to "lay off" because of the extension service attitude: "If you want to love me, you can't love FSA."

3. In 1941, an official in TVA central management initiated a proposal that the Authority undertake a general understanding with Farm Security Administration whereby joint effort might be concentrated on the problems of migratory farm labor, surplus farm population, and the reaching of progressively lower economic strata of farm families. It was suggested by this official that the approach of TVA and FSA were the same. This proposal was rejected by the TVA agriculturists, in a vein which called into question the validity of the FSA enterprise as a whole and insisted that the methods of the two agencies in agricultural matters were quite dissimilar, thus precluding any general understanding. As in other instances of this kind, the agriculturists pointed out that FSA clients were free, as individuals, to participate in the TVA extension service program. This formula, which shunts the problem of interagency coöperation from the level of organizational relationships to that of participation by individual clients, avoids the organizational implications of coöperation and in effect does not constitute meaningful coöperation. The planned programming of a joint enterprise is a far cry from a permitted overlap of clienteles.[32]

4. While the FSA was carefully insulated from the major phases of the TVA agricultural program, a point of official contact between the two agencies was established in 1937. This involved participation of the FSA in the work of aiding some of the families displaced by the TVA reservoirs. The Relocation Section of the Reservoir Property Management Department did not share the anti-FSA bias of the agriculturists and was willing to make recommendations to central management favorable to official coöperation with FSA. In connection with this work, a representative of FSA was provided with office space at TVA headquarters to function as an informational coördinator. The agriculturists objected to this arrangement from the beginning, demanding that any arrangement with FSA should include the land-grant colleges as a party.[33] However, the FSA representative remained. But upon the occasion {169} of a routine resignation of the designee, the agriculturists moved to thwart the appointment of a successor. This was accomplished in a way which revealed the high order of self-consciousness which the agriculturists brought to bear upon all matters of organizational relationship. The Chief Conservation Engineer, acting as spokesman for the agriculturists,

> . . . advised in considerable detail of the efforts that were now and for the past few months had been under way in the Department of Agriculture at Washington to develop a procedure under which the programs and activities of the action agencies like Farm Security could be coördinated with and channeled through the extension agencies in each state. He advised that an arrangement was being or had been developed under which a committee would be established in each state, with the State Extension Director serving as chairman of the committee. The Bureau of Agricultural Economics was to serve as the coördinating agency and would furnish the secretary to each state committee. [The Chief Conservation Engineer] said he would like to see the Authority coöperate in the development and maintenance of the over-all plan rather than to ask a special arrangement be provided through which to carry on the coöperative relationships.[34]

In this way, the agriculturists introduced matters of broad national policy into the determination of specific problems. It would be impossible to understand this move without the knowledge, developed above, of the commitment of the TVA agriculturists to the extension service wing of the Department of Agriculture. The viewpoint of the Chief Conservation Engineer was accepted, and the TVA request for an FSA representative was withdrawn. It seems doubtful that FSA appreciated the Authority's concern lest it "cause the FSA to deviate from the policy and procedure recently developed in order to make its facileties available to displaced families."

THE SOIL CONSERVATION SERVICE CRISIS

The basic pattern of administrative coöperation envisioned by the TVA agricultural leadership found its expression in a three-way memorandum of understanding uniting the Department of Agriculture, the TVA, and the land-grant colleges of the seven Valley states.[35] This agreement established a Correlating Committee, and included the fundamental provisions that all proposals for coördinated activity on the {170} part of any of the parties would be channeled through the joint committee. Viewed in its broader context, the agreement and the machinery it established functioned to unite the old-line elements in agricultural administration against the new agencies which, it was feared, might overwhelm and displace the established institutions. At the time of the execution of the memorandum of understanding, in 1934, the representatives of the several agencies involved shared the same basic outlook, so that unity was real rather than merely verbal. However, as the administrative offspring of the New Deal in agriculture gained strength and stability, there was observed a weakening of the support given by the Department of Agriculture to the three-way understanding in the Tennessee Valley area.

The push and pull within the USDA was reflected in the changing attitudes of the Department toward earlier administrative arrangements and, more concretely, in the ascent of new personnel, with divergent outlooks, to high posts within the Department. Thus, as the Department changed, its role as party to the memorandum of understanding necessarily changed. The earlier informal meaning of the agreement, which presumed a unified outlook on the part of the participants, became a source of embarrassment to newly powerful elements within the Department. This change was reflected in the membership of the Correlating Committee when Milton S. Eisenhower, Land-Use Coördinator, became the representative of the Department. Eisenhower tended to represent the "new" Department and was not wholly sympathetic with the outlook and aims of the TVA agriculturists. As a result, the real content of the TVA-USDA-colleges pattern of coöperation was seriously altered, despite the continued existence of the written memorandum and the formal procedures of coöperation. The deterioration of the old relationship, from the standpoint of the TVA officials, was highlighted in 1941 by a critical situation which almost resulted in an open break between the Department and the Authority, and which caused the latter to make explicit the long-run organizational consequences which a change in the established procedures might invoke. This crisis situation (or "casual breakdown")[36] casts into bold relief the implications of the grass-roots policy for the place of the Authority within the federal administrative system. {171}

Among the new agencies which departed significantly from the administrative method of the agricultural extension service is the Soil Conservation Service. This organization was established in 1933 within the Department of the Interior as the Soil Erosion Service, under the authority of the soil erosion control provisions of the National Industrial Recovery Act. Under the leadership of a dynamic and evangelistic former member of the Bureau of Chemistry and Soils, Hugh H. Bennett, the agency grew to one of the largest of the action organizations within the fields of agriculture and soil conservation. The SES early undertook a radical departure from the methods approved by other farm agencies, especially in the practice of supplying labor and materials to farmers without cost. This direct approach to farmers with "something to offer beside advice and inspiration" placed the extension services in a disadvantageous position and was, moreover, inimical to the basic philosophy of the older organizations, which stressed the need for substantial contribution by the farmers themselves to any conservation program. In addition, the establishment of a new channel of approach to the farmers was a threat to the dominance of the American Farm Bureau Federation, which lobbied in Washington to divert the functions of the SES through the extension service machinery. However, the leadership of the SES was convinced that it had a special mission to perform, requiring a measure of zeal and dedication which could not be preserved if it did not have control over the execution of its soil conservation program. This attitude of dedication to soil conservation was transmitted through the ranks, serving to weld the organization into a coherent group and infusing it with a distinctive idealistic spirit. As time went on, some of the earlier practices of the SES were modified, taking into account that other agencies existed with which stable relations had to be maintained, but its existence as an independent organization, with direct access to the farmers and soil conservation problems, was not compromised.

The Soil Erosion Service was transferred to the Department of Agriculture in March, 1935, and, after approval of the Soil Conservation Act of that year, the SES was designated the Soil Conservation Service. In its new form—but still under Bennett's leadership—the organization entered a period of expansion. With the revamping of the legal foundations of the Agricultural Adjustment Administration in 1936, soil conservation was firmly established as a basic aspect of the action responsibilities of the federal government, and the SCS was given a key role in the administration of the Soil Conservation and Domestic Allotment {172} Act. The original objectives of the SES, centering on the prevention and control of erosion, were considerably broadened in the new responsibilities of the SCS. In 1940, the basic objectives of the Soil Conservation Service were broadly formulated:

> . . . to aid in bringing about desirable physical adjustments in land use with a view to bettering the general welfare, conserving natural resources, establishing a permanent, balanced agriculture and reducing the hazards of floods and siltation.
>
> In attaining these objectives the Service, as an action agency, provides technical and material assistance in determining and making

physical adjustments for the conservation and proper use of soil and water resources and in connection with farm forestry and water facilities development. The Service also engages in sub-marginal land purchase and development.[37]

As the functions of the SCS were extended, and as the intensity of its activities increased, problems of relationship with other agencies having similar functions grew in importance. This situation assumed a peculiar form in the Tennessee Valley, shaped by the special role of the TVA and its Agricultural Relations Department.

During the early days of the Soil Erosion Service and the TVA, some preliminary moves toward joint activity—particularly in relation to a possible soil-erosion experiment station in the Valley—were undertaken, chiefly by representatives of the A. E. Morgan wing of the Authority. Nothing came of these efforts, however, and in 1935 an official statement of the Department of Agriculture defined the relation of the Soil Conservation Service to the TVA. A report submitted to the Secretary of Agriculture on June 5, 1935, by his Committee on Soil Conservation recommended:

> That the Soil Conservation Service not undertake erosion-control work in the area under the jurisdiction of the Tennessee "Valley Authority except as agreed upon by the existing TVA coördinating committee of three representing the Department, the Tennessee Valley Authority, and the seven States in the Tennessee River Basin. This TVA committee has been established to provide coördination of all agricultural activities conducted within that area.[38]

This procedure, which amounted in practice to the virtual exclusion of the SCS from soil conservation work in the Valley, was accepted as basic policy until 1940. Up to that time, the SCS went through a period of great expansion, but refrained from entering those counties of Tennessee, {173} Georgia, North Carolina, Alabama, Kentucky, Mississippi, and Virginia which were defined as within the basin of the Tennessee River. One of the chief means of this expansion was the enactment of soil conservation laws by the state legislatures establishing local soil conservation districts. These districts were authorized to request aid from the Soil Conservation Service, which had a statutory obligation to respond. The state laws were often known as "SCS Bills" and it was generally reported that SCS officials were involved in lobbying activities at the local level, to insure participation by the state in the program. In 1940, after the Kentucky legislature passed such a bill, the SCS is reported to have entered the Tennessee Valley portion of the state, but a really critical problem did not develop until the following year, when it was precipitated by the developing situation in Alabama.

Early in June, 1941, the Northeast and Northwest Alabama Soil Conservation Districts applied to the Soil Conservation Service for inclusion in its regular program of assistance to such districts. This was the first formal issue to be raised envisioning a change in the exclusion order of 1935 which restrained the SCS from working within the Valley area. The matter was placed before the Correlating Committee,[39] resulting in a two-year period of negotiation and discussion which made explicit the underlying organizational assumptions which the TVA had made a part of its grass-roots approach in agriculture.

The significance of the SCS invasion of the Tennessee Valley area was immediately apparent to the TVA agricultural leadership. In July, 1941, within a month of the application of the Alabama districts for SCS aid, the TVA Director of Agricultural Relations had laid before his colleagues a detailed exposition of the organizational consequences of the new development. These consequences were seen in the broad context of changing policy and principle within the agricultural organizations of the United States government. Noting the development of "increaseingly wide divergences of point of view" between the Department of Agriculture on the one hand and the Authority and the land-grant colleges on the other, this official called for a general clarification of relationships among the interested agencies, including such farm organizations as the American Farm Bureau Federation, It was suggested that the TVA Board of Directors press for the recognition by the Department of Agriculture of the idea of the Tennessee Valley as "an experimental area for unified development under decentralized {174} administration"; that is, under the primary guidance of the TVA, working with and through the state and local agencies. The issue was phrased in strong terms: the Authority must either capitulate to or aggressively oppose new forces and principles which necessarily would lead to the defeat of the Authority's multipurpose function. Moreover, the actual disturbance of existing relationships among agricultural agencies operating in the Tennessee Valley was seen to have reached such a critical stage as virtually to force upon the TVA the necessity to come to some over-all decision whether it would acquiesce in the new changes of policy or refuse to do so. In either case, far-reaching consequences would have to be predicted and faced.

The TVA agricultural program (so the argument ran) is rooted in a basic decision that the resources of the land-grant college system within the Valley be utilized to the full, and no independent TVA field force be established. At the time this decision was made, this reflected the dominant administrative philosophy of the U. S. Department of Agriculture. But gradually the TVA agriculturists noted a change in the attitude of the Department. It was observed that the Department's original support of the program based on the memorandum of 1934 weakened as its newer action agencies increased in strength and influence. At the same time, disagreements arose between the Department and the land-grant colleges with respect to the allocation of programmatic functions and responsibilities. Concomitantly with changes in viewpoint, a rapid turnover of administrators within the Department relegated the three-way Valley agreement to the limbo of the forgotten and the unenforced.

The new proposal of the Department, that the SCS enter into agreements with the Northeast and Northwest Alabama Soil Conservation Districts, was not acceptable to the TVA officials. They maintained that such a move would involve the Authority in relationships not envisioned by the memorandum of understanding. In effect, this could only mean that the TVA so interpreted its obligations as to force it to insist that the extension service had an exclusive right to execute an agricultural program within the Valley area. No attempt had been made by the TVA to mobilize the resources of the new federal agencies in agriculture or to fit them into a regional agricultural program. Nor could the Authority do so and yet insist upon the exclusive organizational legitimacy of the extension services. Moreover, the TVA was in this instance objectively carrying out the program of the American Farm Bureau Federation which, as we have already noted, was likewise {175} opposed to the Soil Conservation Service. Nor did the TVA conceive of the SCS situation as a minor or passing affair: It was considered evident that the new approach of the Department of Agriculture, as reflected in its support of the SCS proposal to work in northern Alabama, constituted a threat to the TVA's basic concept of a decentralized regional agency. Consequently, the issue at hand could not be considered in isolation, but rather in terms of its general meaning for the course of the entire regional program.

The TVA agriculturists flatly believed that, when the Alabama issue was brought before the Correlating Committee, the Department of Agriculture's proposal should be rejected. But the consequences which such a course of action involved were not, from the viewpoint of TVA, altogether desirable. If the Authority were to reject the Department's proposal, the latter might simply withdraw from the memorandum of understanding. Then the Department might proceed to muster all its resources for an all-out program within the Valley area, possibly inducing personnel on the TVA-extension test-demonstration program to leave for better-paying positions; it might even stimulate and sustain whatever local opposition there might be to the Authority's electric power policy. Pressure to forego earlier loyalties to the TVA might be brought to bear upon state institutions under threat that flexibility in the use of federal funds would be curtailed. With this line of reasoning, the TVA agriculturists implied that the TVA might lose an effective base of support for its whole program, especially in electric power, unless the problem of relationships in agriculture were solved in such a way as to preserve the original pattern. More generally, it was suggested that the Department of Agriculture might use the Authority's rejection of its proposal in Alabama to support the contention that the establishment of regional agencies such as TVA is administratively undesirable, especially in relation to the possible congressional approval for the inclusion of the Cumberland River basin within the jurisdiction of the TVA.[40] Or, further, the Department might withdraw from its agreement to utilize TVA phosphates in the AAA program, resulting in serious embarrassment to the Authority. Finally, if, in the face of vigorous action by the Department in the Valley, the TVA should continue to press its own program, the result would be widespread confusion and duplication which would not long be tolerated by Congress. In the event of congressional action against the TVA on this issue, the {176} dismemberment of the Authority, long desired by its enemies, would be on the road to accomplishment.

But the Authority's dilemma in this crisis was completed in that, should decision be made to acquiesce in the proposals of the Department, the consequences might not vary greatly from those of a decision to oppose them. The alternative was phrased in these terms:

> If the Department should intensify its programs within the Tennessee Valley under its own direct methods, which the coöperating state institutions would be forced to accept just as they have been forced to do in the areas outside the Tennessee Valley, the Authority would be forced, with the allegiance of the institutions to TVA thus alienated, either to *withdraw from the agricultural field or to initiate its own program directly.* To withdraw means the dismemberment of TVA, the very thing which the enemies of TVA have advocated for many years. To proceed directly with its own program would violate the principles for which the Authority has long stood. It would increase enormously the expenditures in the area through the prosecution of duplicating programs. This the Congress would not countenance, and properly so.

In fact, it seems to have been realized within the Authority that the agency could not hope to escape completely from the pressure of these harsh alternatives, and that the situation could be saved if at all only by avoiding a head-on clash with the Department of Agriculture. Some reconciliation was necessary which would preserve the form of the old relationships and limit as far as possible the actual extension of SCS operations in the Valley. However, while such an administrative clash would have to be avoided, it was considered possible that the Authority might muster its political resources in order to effect a favorable resolution of the situation. This orientation was expressed thus:

> It is imperative, therefore, that the Authority and the Department come to terms and that the Department return to its allegiance to the memorandum of understanding. The first step in reconciliation of ideas and objectives would seem to be for the Board of Directors of the Authority, either alone or with representatives of the land-grant colleges, to approach the Secretary of Agriculture to determine: 1) His understanding of the relationships of the Department to the Authority; and 2) His position with reference to the existing memorandum of understanding. In the event he is found to be unfavorable to the principles of the memorandum, this fact should be known to the Authority and to the coöperating institutions, who would then inform agricultural leaders and farm organizations throughout, the country and particularly in the Tennessee Valley, so they then might inform

their state representatives in Congress and request that the matter be resolved by the President of the U. S. or in the Halls of Congress.

It is not known whether such pressure was actually brought to bear at the highest political levels, but it is known that a direct effort was made {177} to enlist the leadership of the American Farm Bureau Federation on the side of the Authority in this situation. Moreover, a special meeting of the presidents of the land-grant colleges of the Valley states was held in June, 1942, with the participation of Director H. A. Morgan of the TVA Board. This meeting was called expressly to show the solidarity of the land-grant institutions with the TVA and its approach. The TVA found itself in a difficult position with relation to the colleges. Its own commitment to the grass-roots method had made the continued coöperation of the colleges indispensable, but the colleges could not restrict their consideration to the effect of new patterns upon the TVA as an organization. This was exhibited in the dilemma of the Director of the Alabama extension service who was caught between his sympathy for the TVA and his inability to refuse the concrete help which SCS was in a position to give to Alabama farmers. He could not accede to pressure from the TVA to oppose the entry of the SCS into the Valley portion of the state, particularly as that portion represented only a small part of the state's total area, a consideration which must have been important for all of the seven Valley states except Tennessee. However, in the course of the meeting of college presidents in June, 1942, the TVA was able to gain (or perhaps salvage) the following commitments from the institutions:

> 1. That, although they recognized the administrative obligation of the Authority to uphold the regional program, they asserted the concern of the institutions in their own right in a program of unified development as a means of fulfilling their institutional obligations....
>
> 2. That the responsibilities of the Department, the institutions, and the TVA with respect to the regional program were so clear and binding that, regardless of the merits of the North Alabama issue, none of the parties should withdraw from the program. Failure to coöperate would endanger the success of the national experiment which the Congress and the Administration had set up in the public interest.

The TVA was wholly dependent upon the continued coöperation of the state institutions in its agricultural program, so it was imperative to mobilize their support for the continuation of the memorandum of understanding. It was well understood by the TVA officials that the Authority could not hope to work out its relationships with the land-grant colleges independently of the Department of Agriculture. Hence the continued existence of a three-way agreement was considered indispensable.

At a meeting of the Correlating Committee on September 4, 1941, the general situation was reviewed, but far from any agreement being {178} reached, there ensued a difference of opinion over even the conclusions reached at the meeting itself. In any case, immediately upon his return to Washington, Land-Use Coördinator Milton S. Eisenhower extended his approval, on behalf of the Secretary of Agriculture, to the general agreements executed between the Northeast and Northwest Alabama Soil Conservation Districts and the USDA's Soil Conservation Service. Eisenhower took the position that conditions had changed since the exclusion order of 1935 and necessitated a revision of the earlier procedures. The Department officially recalled that the 1935 functions of SCS consisted largely of research, employment of relief labor, and the management of watershed demonstration projects in a relatively small number of areas. The demonstration approach of the SCS had now changed to one of extensive assistance to the states, with the objective of reaching the greatest possible number of farmers. Congress had appropriated funds for this work, on the assumption that no administrative barriers would be placed in the way of assistance to all farmers eligible under the law and requesting it through the local soil conservation districts. Under pressure of these conditions, it was felt that the Department could no longer feel justified in excluding the SCS from operation within the Tennessee Valley.

At the same time, the Department was willing to compromise in relation to the TVA program to the extent of enjoining the SCS to take into account the special circumstances within the Tennessee Valley: (1) The SCS was to "refrain from carrying on promotional work aimed specifically at the organization of soil conservation districts within the Valley"; (2) the SCS was to avoid working on any farm in the TVA-extension service test-demonstration program "without first obtaining the concurrence of the local officer in charge of such TVA demonstrations";[41] (3) services to the soil conservation districts were to be carried on in close coöperation with extension service personnel working in the districts.

In the course of the negotiations, the Department stressed that in fact it had not generally operated through the Correlating Committee within the Valley area. With the exception of SCS, the action programs of the Department had been carried on directly within the Tennessee Valley just as in other parts of the country. These programs included those of the Agricultural Adjustment Administration, the Farm Security Administration, the Farm Credit Administration, and the Rural {179} Electrification Administration. This was often ignored by the TVA officials, who persisted in the assertion that the SCS crisis revolved about a generally new approach by the USDA in the Valley. In fact, the memorandum of understanding did not have the controlling force which the TVA agriculturists wished to give it, though it was used as a weapon in support of TVA's constituency commitments whenever possible.

The Department did not recede from its position that the exclusion order of 1935 with respect to SCS would have to be abandoned. In 1942, after a protest by the chairman of the Correlating Committee, the

Secretary of Agriculture served notice that the memorandum of understanding would be terminated. However—probably after further negotiations at the top executive level among the agencies involved—a new memorandum of understanding was signed in September, 1942, which included the following statement as the only significant change in the earlier agreement:

> It is understood that neither this memorandum of understanding nor the Correlating Committee provided for herein, has administrative jurisdiction which would prevent any signatory agency or institution from carrying out the provisions of legislation for which said signatory is responsible.[42]

This new language permitted the Department to carry forward its SCS activities without challenge by the Correlating Committee, since it was held that upon request assistance to the soil conservation districts was required by law. At the same time, the over-all framework of the agreement—now even formally reduced to little in the way of regional integration beyond the relation of TVA and the colleges—was maintained as a gesture to the continuity of the old relationships.

In this chapter we have examined one phase of the consequences of the exercise of discretionary power at the grass roots. The commitment of the TVA agriculturists to the land-grant college system involved the agency as a whole in discordant relationships with the federal administration. Intervention in the Valley occurred during a period of general controversy in the field of agricultural administration, heightening the self-consciousness of all participants. Hence this analysis provides a detailed illustration of the basis for the Authority's failure to adjust itself to the going federal administrative structure.

---

[1] See above, pp. 64-68.

[2] See above, p. 73.

[3] See above, chaps. iii, iv.

[4] See Russell Lord's informal history, *The Agrarian Revival* (New York: American Association for Adult Education, 1939). Also Gladys Baker, *The County Agent* (Chicago: University of Chicago Press, 1939). Earlier, and more detailed, is the two-volume *Survey of Land-Grant Colleges and Universities*, Office of Education, U. S. Department of the Interior (Washington: Government Printing Office, 1930).

[5] See above, pp. 117-124.

[6] See above, p. 121.

[7]
The Farm Bureau contented itself with ignoring the FSA in its appropriation recommendations before 1940. Later, especially in 1942 and 1943 under the impetus of an economy drive in Congress, it conducted an all-out attack. The conclusion seems inescapable that, had the TVA undertaken an agricultural program along the lines of the FSA, the formidable opposition of the Farm Bureau would have had to be faced and, conceivably, might have turned the scale in favor of those who wished to dismember the Authority.

[8]
Quoted by Carlyle Hodgkin, "Agriculture's Got Troubles," the second of a series in *Successful Farming* (April, 1943), p. 26.

[9]
Oren Stephens, "FSA Fights for Its Life," *Harper's* (April, 1943), p. 478.

[10]
Oscar Johnston, President, National Cotton Council of America. *Hearings*, House Committee on Appropriations, Agriculture Department Appropriation Bill for 1944, 78th Cong., 1st sess. (1943), p. 1619.

[11]
Donald Kirkpatrick, General Counsel of American Farm Bureau Federation. *Hearings*, Joint Committee on Reduction of Nonessential Federal Expenditures, 77th Cong., 2d sess. (1942), p. 815.

[12]
Ibid., p. 742.

[13]
See J. M. Gaus and L. O. Wolcott, *Public Administration and the U. S. Department of Agriculture* (Chicago: Public Administration Service, 1940), p. 195: "The Administration of the [Agricultural Adjustment] program also had to meet the pressures of commodity groups, the organization and support of which had long sustained a demand for national farm program. The Administration had to cope with the pressures of groups that, since they represented primarily the owner class, pressed for policies favorable to themselves regardless of the interests of tenants and sharecroppers. In making these demands effective, the AAA came into conflict with a group in the Department that, to compensate for the injustices, became a motivating factor in the development of the F.S.A."

[14]
See above, pp. 119-121.

[15]
Editorial, *National Farm Holiday News* (January 1, 1937).

[16]
See Clifford V. Gregory, "The American Farm Bureau Federation and the AAA," *Annals of the American Academy of Political and Social Science*, 179 (May, 1935), 152-156.

[17]
See Baker, *op. cit.*, pp. 70 ff., for details of aid derived by the extension service from AAA funds.

[18]
See Lord, *op. cit.*, p. 163.

[19]
See testimony of E. A. O'Neal, *Hearings*, Joint Committee on Reduction of Nonessential Federal Expenditures, 77th Cong., 2d sess. (1942), pp. 754-755.

[20] The underlying hostilities were of course well understood in agricultural circles. One reporter wrote in March, 1943: "In the beginning the American Farm Bureau Federation was the great champion of AAA. It fought twenty years for the principle of 'parity price.' It battled valiantly for the farm program in the halls of Congress. It steered the new program through its early wobbly days. But now that AAA has grown big and powerful, with a committee of five farmers in every precinct in the U. S., the Farm Bureau begins to fear it as a *rival* farm organization." Carlyle Hodgkin, *Successful Farming* (March, 1943), p. 54.

[21] This issue of the *Tennessee Farm Bureau News* published lengthy quotations from speeches of AAA officials attacking the Farm Bureau-extension service bloc.

[22] *The Nation's Agriculture* (organ of the American Farm Bureau Federation) (March, 1942), p. 2.

[23] See *The Role of the Land-Grant College in Governmental Agricultural Programs*, Iowa State College Bulletin (Ames, Ia.: June, 1938).

[24] See Lord, *op. cit.*, p. 162.

[25] Milton S. Eisenhower, "Who Should Be Responsible in the Development of an Agricultural Planning Program?" an address by the Land-Use Coördinator and Director of Information, U. S. Department of Agriculture, at the meeting of the Association of Land-Grant Colleges and Universities, Chicago, November 14, 1938 (mim.).

[26] See Stephens, *op. cit.*, p. 481: "There arose consequently a conflict between two agricultural groups, and indeed, between two groups within the USDA. To be specific, this conflict was between the Extension Service, the AAA and the Farm Bureau on one side, and the FSA and the Farmers Union on the other. Basically, this is the line-up as it exists today [1943]."

[27] See Gaus and Wolcott, *op. cit.*, p. 66.

[28] The Agricultural Adjustment Administration has played an important role in the distribution of TVA fertilizers, but apart from the purchase of fertilizer *en bloc* there has been little in the way of a close organizational relationship. It appears likely that the decided advantage offered by AAA in providing a convenient outlet for surplus fertilizer materials outweighed any considerations of interagency controversy.

[29] See above, pp. 141-145.

[30] These considerations are general, and apply as well, of course, to the relations between political parties, for example, and other nongovernmental groups of all kinds.

[31] See above, p. 112.

[32] This is the same sort of problem as a "united front" in politics. If a joint political meeting is proposed by one organization to another, it may be

suggested by one—if it really wishes to go ahead on its own—that "there is no need for organizational alignments, anyone may come to the meeting." Such a reply will not be accepted in good part by officials who must pay attention to the problem of party recognition, joint leadership, etc.

[33] The tactic of introducing the land-grant colleges as a necessary party, with which the TVA agriculturists countered any suggestion that the Authority enter into coöperative arrangements with FSA, was often repeated, and served as an effective way of hinting that the FSA was "not wanted," or at least, if there had been (as there was not) a serious challenge, as a delaying maneuver.

[34] At a meeting, March 3, 1939, which included the General Manager, the Chief Conservation Engineer, and representatives of the TVA Reservoir Property Management Department.

[35] See above, pp. 95-98.

[36] The crisis situation in the individual presents "a short and dramatic dislocation of his usual relationships with any given social institution or social pattern," and "reveals problems which exist in ordinary life and which, in ordinary life, do not carry sufficient emotional reaction to lift them into the field of awareness." See J. S. Plant, *Personality and the Cultural Pattern* (New York: The Commonwealth Fund, 1937), pp. 58-59.

[37] H. H. Bennett, Chief, Soil Conservation Service, and M. L. Wilson, Director of Extension Work, "A policy Statement on Relationships between the Soil Conservation Service and the Extension Service," March 29, 1940 (mim.).

[38] U. S. Department of Agriculture, "Report of the Secretary's Committee on Soil Conservation," June 5, 1935 (mim.). This report was approved by Secretary Wallace on the following day.

[39] Dean Thomas P. Cooper, University of Kentucky Extension Director; Director J. C. McAmis, Agricultural Relations Department, TVA; Land-Use Coördinator Milton S. Eisenhower, U. S. Department of Agriculture.

[40] It will be recalled (p. 78) that this issue did in fact arise during 1941, and that the Department of Agriculture took its position on organizational grounds.

[41] It had earlier been charged that SCS activities in Kentucky had involved a conscious effort on the part of SCS officials to "sign up" TVA demonstration farms.

[42] "Statement and Memorandum of Understanding between U. S. Department of Agriculture, Tennessee Valley Authority, and Agricultural Colleges of Alabama, Georgia, Kentucky, Mississippi, North Carolina, Tennessee and Virginia." Effective October 16, 1942. Signed by the Presidents of the Institutions, the Chairman of TVA, and the Secretary of Agriculture (mim.).

CHAPTER VI

LAND-USE POLICY AND THE CHARACTER OF TVA

The purchase of a minimum amount of land around TVA reservoirs would provide opportunity for private enterprise to exercise its initiative in the development of the waterfront free from the stifling effect of public ownership and in accord with the public demand.

THE TVA AGRICULTURISTS (1941)

THERE IS A VAGUE and ill-defined quality which, unacknowledged and often poorly understood, represents a fundamental prize in organizational controversy. This is the evolving character of the organization as a whole. What are we? What shall we become? With whom shall we be identified? Where are our roots? These questions, and others like them, are the special responsibility of statesmen, of those who look beyond the immediate context of current issues to their larger implications for the future role and meaning of the group. To pose these questions is to seek more than the technical articulation of resources, methods, and objectives as these are defined in a formal program or statute. To reflect upon such long-run implications is to seek the indirect consequences of day-to-day behavior for those fundamental ideals and commitments which serve as the foundation for loyalty and effort.

"Consequences," writes Dewey, "include effects upon character, upon confirming and weakening habits, as well as tangibly obvious results." And in an earlier passage from the same work: "Character is the interpenetration of habits. If each habit existed in an insulated compartment and operated without affecting or being affected by others, character would not exist. That is, conduct would lack unity being only a juxtaposition of disconnected reactions to separated situations."[1] These considerations from the social psychology of the individual provide us with tools for organizational analysis. Organizations, like individuals, strive for a unified pattern of response. This integration will define in advance the general attitudes of personnel to specific problems as they arise. This means that there will be pressure within the organization, from below as well as from above, for unity in outlook. As unity is approximated, the character of the organization becomes defined. In this way, the conditions under which individuals may "live {182} together" in the organization are established, and a selective process is generated which forces out those who cannot identify themselves with the evolving generalized character of the organization. The evolving character, or generalized system of responses, will be derived in large measure from the consequences of day-to-day decision and behavior for general patterns of integration.

An examination of the logic of this development provides us with a theoretical link between the concept of the character of an organization and that of administrative discretion as discussed above.[2] The act of discretion permits the administrator to introduce considerations tangential to the formal or stated objectives of the organization. At the same time, by the accretion and integration of modes of response, the officialdom is able to invest the organization with a special character. This special character tends in turn to be crystallized through the preservation of custom and precedent. It is further reinforced by the selective process which rejects those members who cannot fit in, and shapes the personal orientation of those who remain or who are recruited.

In a situation charged with conflict, the process of discretion will be subjected to close scrutiny, and the quality of administrative decision will tend to be infused with a high degree of self-consciousness. The scrutiny of the opposition and the self-consciousness of the leadership will alike center upon the question of commitment. What attitudes and what symbols are commanding the loyalties of the staff? What precedents are being established? What alliances are being made? Such issues will be uppermost in the minds of leading individuals during periods when the evolution of the character of an organization is not yet settled. The possibility of stating that some given line of action is the "settled policy of this organization" is one of the strategic objectives in such a controversy. Or, in a field somewhat oblique to questions of policy, there may be conflict over the "heroes" of the group. Laudatory references to a set of individuals as the "fathers" of the organization's policy and outlook may help to define the accepted antecedents of the group; as a result, a whole series of doctrinal commitments are inferred from those antecedents, though not necessarily formally included in the program or objectives of the organization. In such cases, controversy may occur over the question of whether some individual's memory is to be celebrated in the official newspaper or bulletins, or whether certain slogans and symbols, traditionally associated with one general tendency or another, will be included. Conflicts over apparently {183} minor matters of this sort are typically aspects of the struggle to determine the character of an organization.

The internal organizational pressures which drive toward a unified outlook and systematized behavior receive their content, or substantive reference, from the play of interests and the flow of ideas which characterize the organization's social environment. In this way, the internal process of character formation—though generated by needs which may be referrable primarily to the organization as such—

comes to be stamped with the typical hallmarks of its own historical period. The general commitments and attitudes of the organization (i.e., its character) will tend to crystallize around value problems which are current in the environment. For an organization whose discretionary power on social questions is broad, there is pressure to make a choice among the "historical alternatives" that are available. Once that choice is made, the organization will tend to reflect in its own character the general sentiments with which it has become aligned.

The struggle over the outcome of this process may extend over a long period, and may be compromised from time to time, but the stake is always all important. The attempt to define an organization's character cannot be divorced from the struggle for leadership or from the possibility of internal convulsion. It is precisely in the struggle over an evolving organizational character that a given leadership having certain personal qualities most easily becomes the receptacle of a social ideal. As such, that leadership—incumbent or proposed—is conceived as indispensable to the goal of stamping the organization as a whole with the desired ideal. A leadership can become indispensable when it has convinced itself and its constituents that some alternative elite cannot be trusted with the exercise of character-defining discretionary powers; whereas the possibility of adequate replacement in the execution of formal executive functions is not normally in doubt.

These preliminary remarks permit us to round out the logic of our thesis, which may be stated in these summary terms:

1. The theory of grass-roots administration as reflected in the idea of "working with and through local institutions" remains unanalyzed prior to an explication of its concrete application and consequences.

2. In the major field of the TVA agricultural program, the exercise of discretion in applying the grass-roots concept initiated a constituency relationship involving the land-grant college system on the one hand and the Agricultural Relations Department of TVA on the other.

3. The constituency relation may be viewed as a mechanism of the {184} informal coöptation of grass-roots elements into the policy-determining structure of the TVA.

4. The impingement of the administrative constituency upon policy, operating through its organized representatives within the Authority, takes advantage of the special prerogatives of the latter group. These prerogatives derive from its status as an integral part of the formal structure of the Authority.

5. The prerogatives of the Agricultural Relations Department include the exercise of discretion within its formally assigned subject-matter field as well as the exertion of pressure upon the evolution of general policy within the TVA as a whole.

6. Especially in a situation marked by controversy, the selective consequences of the pressure exercised by the TVA agriculturists sig-

nificantly affects the role of the TVA as a whole within the federal system.

7. As major character-defining issues come to the fore, this pressure influences the evolving character of the Authority as a whole.

8. Consequently, our search for the meaning and significance of the grass-roots policy verifies the hypothesis that the channeling of administrative responsibilities through previously existing institutions result in consequences for the role and character of the initiator which were not anticipated by the formal theory or doctrine. We also learn something about the mechanics of this process.

These points are demonstrable on the basis of the empirical analysis presented in this study. However, it should be mentioned that there are specific matters of fact, for which adequate direct evidence is not available, which are considered significant by informed participants. These include (1) the special relation of the Agricultural Relations Department of TVA to Dr. H. A. Morgan, one of the three members of the TVA Board, with the implication that the TVA agriculturists had in effect a direct representative on the Board of Directors—a privilege enjoyed by no other single department in the Authority; and (2) the political importance of the extension service machinery, spread throughout the Valley, as enhancing the power of the agriculturists within the Authority—again the only department within TVA having the potential power of wielding an organized apparatus in the interests of policy questions which it might consider crucial.[3] These considerations, {185} among others, help to explain the relatively dominant position of the agricultural group, resulting in a degree of influence upon policy which is not seen from an inspection of the formal structure of the Authority.

The thesis thus stated concentrates attention upon organizational dynamics. It is necessary to bear in mind, however, that the evolution of policy may be traced to broader and more general factors than the specific pressures exerted by groups within and around a given organization. In a sense, the forces generated by the process of organization per se may be thought of as the means by which the pressing but more general imperatives of the particular historical period are given effect. Such motives as prestige and survival may adequately impel action; but in general the need to rally forces broader than the small group for whom these motives are effective will make necessary an appeal based on moral or political principles which can be defended on their own level. In this way, the organizational struggle is provided with doctrinal content and a socially acceptable arena. But the price of that strategy involves a commitment to a set of ideas or interests. Hence those ideas and interests are provided with a means of intervention, a driving force which they may not be able by themselves to generate, but which, once generated, can carry them along with it.

To this qualification we may add another: changing historical conditions may seriously affect the choices available to discretionary power. In an administrative agency, controversies over policy may become academic overnight if statutory changes occur. Or the range of choice may be gradually restricted by changing economic conditions—as from a period of depression to one of prosperity, or of relative plenty to relative scarcity—as well as by shifts in the political climate which may make certain choices less expedient at one time than another. As these changes occur, the relative strength of forces within the organization may vary, reflecting needed reorientations. As the new realignment takes place, policy will change, but it might be rash not to look beyond the inner-organizational controversies for the cause of the change. In general, it is suggested that these considerations—which link specific organizational pressures to the more general imperatives and forces of the time—will temper any tendency which may arise from a concentration upon organizational dynamics to ignore the goals and demands provided by a particular historical period.[4] {186}

TVA AND PUBLIC LANDS

We may now turn to a case history in the evolution of administrative policy which carries forward our search for the consequences of the grass-roots doctrine, and at the same time exemplifies the general principles which have been set forth above. This policy is within the province of discretion of the Board of Directors of the TVA. It defines the extent to which the Authority engages in a program of land acquisition for public purposes incidental to a major project: the construction of a series of dams and the consequent impounding of large bodies of water along the course of the Tennessee River and its tributaries.

The term "incidental" used here is of special significance. Both the theoretical and the practical interest in this record stem from the fact that this is an area of discretion available to the Board. In 1941, the general manager of TVA stated that the problem of determining the boundary ("taking line") around the reservoirs, within which land was to be purchased, represented a major area of difficult administrative coördination within the agency. But this kind of difficulty can arise within a field of decision only when the latter is not rigidly determined or exhausted by technological considerations. The attainment of statutory objectives, when a measure of leeway or choice within even a limited sphere is allowed, opens the door to differences as to incidental or tangential goals. These may be assigned a formal collateral status and thus be given due consideration in the execution of general policy. As we have suggested, controversy over these collateral goals may in effect place the character of the organization at stake.

As with other "general-welfare" or developmental objectives, the legal power of the TVA to acquire public land for general purposes is severely limited; here, too, the context of the Authority's origin and the broad conceptions of its proponents express a fuller set of objectives than does the Act itself. President Roosevelt's message to the Congress of April 10, 1933, had a prophetic ring:

> It is clear that the Muscle Shoals development is but a small part of the potential public usefulness of the entire Tennessee River. Such use, if envisioned in its entirety, transcends mere power development: it enters the wide fields of flood control, soil erosion, afforestation, *elimination from agricultural use of marginal lands*, and distribution and diversification of industry. In short, this power development of war days leads logically to national planning for a complete river watershed involving many states and the future lives and welfare of millions. It touches and gives life to all forms of human concerns.[5] {187}

The preamble to the Act was likewise broadly phrased:

> To improve the navigability and to provide for the flood control of the Tennessee River; *to provide for reforestation and the proper use of marginal lands in the Tennessee Valley*; to provide for the agricultural and industrial development of said valley; to provide for the national defense....[6]

Despite these broad affirmations of purpose with respect to a public land policy, the provisions of the Act which followed the enacting clause neither established machinery nor delegated authority which would give effect to those ambitions. The Authority was assigned power to acquire real estate "for the construction of dams, reservoirs, transmission lines, power houses, and other structures, and navigation projects at any point along the Tennessee River, or any of its tributaries," and a condemnation procedure was established. In addition, Sections 22 and 23 of the Act authorized studies, experiments, and demonstrations, as well as recommendations to Congress which would aid in the proper use, conservation, and development of the natural resources of the Tennessee River drainage basin and adjoining areas. However, these general-welfare provisions of the Act are regarded as among its weakest parts, and the management of the agency has steadily curtailed efforts to base substantial parts of the program upon them.

Owing to these weak statutory delegations, the Authority has followed the policy of executing what it considered to be its general-welfare responsibilities in an oblique way: essentially by investing the administration of those programs which did have a sound legal foundation with certain ideals of public policy. Thus the TVA accepted responsibility for the increased malarial hazard which resulted from the creation of new lakes and backwater areas; the creation of construction villages was conceived as generating responsibility for the provision of social services, including library facilities; the effect of newly impounded reservoirs upon local communities justified efforts to aid in the establishment of town planning services; and the need to protect the public investment in dams and reservoirs was linked to a general interest in soil conservation on the basis of the theory that siltage could be effectively forestalled by minimizing soil erosion in the drainage basin. These considerations, and many others, effectively distinguish the Authority's administrative methods from those of relatively irresponsible private corporations. At the same time, however, the discretionary element in administration is vastly enlarged. To the extent that such tangential responsibilities are accepted, without clear-cut {188} definitions of policy in the statute, the possibilities of variant interpretations are present. "The ideals of public policy" may then depend upon the specific interests and backgrounds of those actually charged with the day-to-day administration.

In an agency such as TVA, land policy—in acquisition and in management—is a criterion to discern the generalized attitudes of responsible administrators on significant issues of social policy. The purpose for which TVA has acquired tracts of land in the course of its general operations have been listed by a TVA official as follows:

> *Essential or unavoidable acquisition*: to acquire reservoir areas necessary for impoundment or flooding; to acquire those areas necessary for the construction of dams, power structures, and for the development of related improvements and operations; to control reservoir margin uses not consistent with the purposes of the TVA Act; to consummate transfer of the properties of certain power companies whose holdings and operations have been taken over by the Authority.
>
> *Desirable but not essential acquisition*: to avoid uneconomic severance of property lines through over-purchase of entire tracts; to eliminate excessive costs of reconstruction and relocation of roads, bridges, and other improvements; to provide opportunities for development of public projects such as recreational areas, experimental grounds, game refuges, fish hatcheries and ponds, tree nurseries, etc.; to provide for the protection of the reservoirs from excessive silting and surface run-off through public control of adjacent lands.

Those purposes listed as "desirable but not essential" and to some extent those listed as "essential or unavoidable" are subject to the discretion of the TVA Board. It is evident, moreover, that the issues involved are of such a nature that long-debated judgments of value must be brought to bear in setting policy within the agency. The role of government in such matters as the provision of recreational facilities and the conservation of natural resources, as more generally the obligation of government to capture for public use the increments of value brought about by technological changes, represents one of the character-defining motifs of the present historical period. This is especially true in those areas of the United States which have been least influenced by such ideologies of the past two generations as the conservationist movement and the ideal of the positive state. Considering the often-remarked persistence of pioneer attitudes toward the land in the Southern states, it is not surprising that the issue of land policy should have been subject to prolonged debate within the TVA organization. It is important for this analysis to discern the mechanics of that debate, and the pressures involved, as these are related to the organizational consequences of the official grass-roots doctrine. {189}

Before tracing the development of the land-purchase policy within TVA, some reference should be made to the viewpoint on this question adopted by public agencies with which the Authority is normally identified. The following statement, from the first issue of the Bureau of Agricultural Economics' *Land Policy Review*, is indicative:

> The first major change of direction in land policy was initiated by the conservation movement which attained momentum in the early years of the twentieth century.... The conservation movement, by instituting public ownership as a means of protecting national economic resources, marked a radical departure from the laissez-faire economic policy of the nineteenth century.... The second great contribution to the stream of land policy . . . has been the development of methods whereby a sound use of land could be achieved on private as well as public lands. [This has marked a change] in our traditional concepts of land tenure.... We are forced to recognize the existence of a social interest in the private ownership of land and to protect that interest where a recalcitrant individual owner fails to do so.[7]

This point of view, which emphasizes the emerging recognition of a public stake in land utilization, is supplemented by the conception of public land acquisition as an instrument of achieving a sound national land-use program. A recent formulation of a general program of land acquisition was undertaken in 1940 by the National Resources Planning Board, and motivated in these terms:

> In the light of present-day conditions, then, private ownership of most of our land resources is generally recognized as a basic tenet. However, public land acquisition of some lands now in private hands is justified, not only for public buildings, wildlife refuges, recreational sites, and for purposes of conservation and restoration of forest resources, but also as a means of assisting in preventing or correcting maladjustments in land use that stand in the way of the fullest and most effective utilization of our national resources.[8]

The NRPB pointed out that programs of public aid to agriculture have apparently relied too much on the potentialities of improved farm management. In its view, however, "there are . . . many lands naturally so unproductive or in such unproductive operating units, that even the extensive public use of adjustment aids now available, plus financial grants, merely enable occupiers to hold on a little longer. Consequently, land acquisition has to-day become one instrument, and a basic one, among many instruments for the effective conservation, development, and wise use of the Nation's resources."[9] The principles embodied in this {190} general policy are widely believed to be those which motivate the TVA, in common with other agencies presumed to stand in the conservationist tradition. However, the Authority's leadership now explicitly states that there is a widespread misconception of the TVA's character in this respect. In fact, the evolution of the Authority's policy on land acquisition and land use is a history of a slow but definite change in outlook, involving a significant and conscious reversal of major policy.

CONSERVATION GOALS IN THE EARLY PERIOD

During the first years of the TVA organization, the "superidealists"—as they came to be known in later days—held many strategic positions within the agency, and attempted to shape its character seemingly consistent with the spirit in which the new enterprise was launched. This group included those most closely associated with Arthur E. Morgan, whose program for social change within the Valley brought him into sharp conflict with Harcourt A. Morgan and the latter's ally on the Board, David E. Lilienthal, within a few months of the founding of the agency. The sharp split within the Board was soon reflected at the lower administrative echelons in hostility and controversy over many issues, among which the question of land policy loomed large both in importance and in persistence.

The first major construction project undertaken by the TVA was the erection of Norris Dam on the Clinch River, one of the large tributaries feeding the Tennessee. The policy which determined the amount of land acquired in the course of this operation marks a distinct phase in the history of the Authority. Acquisitions in the Norris area included an "overpurchase" of approximately 120,000 acres—land lying above the contour which would be reached by the newly created reservoir. Of course these lands, which included vast tracts of forest cover, lay off the main stream of the Tennessee River. It could hardly be expected that acquisition on such a scale would be possible at dam sites along the Tennessee itself. But significant is not so much the extent of the overpurchase at Norris, as the viewpoint within the Authority during that period.

The most aggressive proponent of extensive acquisition was Edward C. M. Richards, Chief Forester of the TVA from 1933 until 1938. Richards was a close associate of Arthur E. Morgan, then chairman of the Authority, and is usually included on the roster of superidealists whose departure is said to have left the agency on firm, realistic ground. He was bitterly opposed from the beginning by the agricultural group. But such factors as the support of the chairman—who was also Chief Engineer {191} and primarily responsible for construction of the dam—in addition to that of the old Land Planning and Housing Division; the general spirit of aggressive idealism which prevailed during the early days; the economic depression which stimulated an interest in the retirement of submarginal lands; and the general character of the

Norris area which seemed to be especially suitable for a large-scale conservation enterprise—all combined to give the position of the Chief Forester a dominance which it would not long hold. Richards attempted to commit the Authority to an interest in the public management of forest lands on an extensive basis, carrying over the experience with such projects which he had gained in Europe. But his concern for the land reached beyond forestry. Thus, in 1934, in the course of a discussion of the evolving land-purchase policy of the Authority, he stated:

> It has been my conviction for a long time that there is a very great need for the TVA not only to discuss some basic policies but to come to some definite conclusions and commit itself to definite policies. Particularly is this true, I believe, in connection with the matter of forest, critical erosion, and marginal land areas within the Valley. All areas falling within these three classifications are of great importance to the Tennessee Valley project because in the first place they are today being used in such a way as to hold down the standard of living of the rural population of the Valley and to increase the soil erosion; and on the other hand, they are not being used in such a way as to constructively build up their potentialities of service in the future.
>
> These three types of land—forest, critical erosion, and marginal—represent a considerable proportion of the total area of the Tennessee Valley. Only in the case of forest lands is any serious attention being given to the care and future development of such areas. And what is being done in the case of forest lands is entirely inadequate, not only at the present time but as far as plans for the future are concerned....
>
> In the case of critical erosion areas and marginal lands, the need for discussion and definite commitment of policy by the Board is in one way even more imperative and pressing than in the case of forest lands, for the reason that for the most part more people are endeavoring to live and earn a livelihood on such areas than on forest lands. The results of such effort are an impoverished rural population and a constantly deteriorating physical condition of the lands themselves.

This point of view, insofar as its goals were concerned, was essentially that of such other governmental agencies as the Soil Erosion Service, the Bureau of Agricultural Economics, and the National Resources Committee.

Richards likewise struck out at the tendency within the Authority to think of soil erosion control as deriving its value solely from the function of preventing silting-in of the dams, rather than as a vital objective for its own sake.[10] In view of later changes in outlook, it is important {192} to make clear this early approach within the TVA to the role of a public agency in land-use policy. Continuing the discussion just quoted, Richards stated:

> I believe that soil erosion control is a part of land use. Proper land use is of vital importance of the TVA to help the population of the

Valley, to stabilize local industry, and to increase the use of electric power. Proper land use calls for the cessation of destructive practices on the one hand and the initiation of constructive productive use on the other. Such proper land use is often impossible financially in the case of the owners of small tracts of land in hilly sections. They must make their own living off of the land they own even though this may tend toward destroying their own property through erosion.

To make possible the proper use of marginal lands, reforestation and erosion control by some public agency is in many cases essential. In order to do this, however, ownership of the land in fee is necessary. This is for the reason that complete control of what is done on the lands necessitates full ownership.[11]

The last two sentences are of special significance, for it is precisely the principle of full ownership in fee simple which the Authority has progressively discarded.

In general, the goals set forth by the first Chief Forester were supported by those within the Authority most closely associated with its planning objectives. The head of the Land Planning and Housing Division, Earle S. Draper, was generally associated with Richards during the early period, though his own professional interest in afforestation as such was more limited. Draper supported the proposals for extensive land acquisition in the Norris area because of "the definite sub-marginality of the land, the extremely low level of income, the lack of opportunity, and the danger of increased settlement to perpetuate low living standards." The professional planners were accustomed to incorporating criteria of social utility as a part of practical judgment, and had long been committed to public ownership and control as a basic instrumentality for avoiding the misuse of marginal areas. The agriculturists within the Authority were deeply committed to an opposing viewpoint, and hotly contested the criteria proposed by the planners and {193} the foresters. But it was not until the resolution of the internal crisis, resulting in the elimination of A. E. Morgan and many of his followers, that the agriculturists were able to exert a disproportionate influence outside their own special province.

THE RESERVOIR PROTECTIVE STRIP

The broad objectives of Chief Forester Richards, quoted above, characterize the first phase of TVA policy on land acquisition. A second phase, for more limited acquisition, is marked by the explicit acceptance by the Board of the policy of establishing a buffer belt of publicly owned land around each reservoir. This policy, analogous to the objective of creating green-belt areas around urban communities, was consistent with the professional outlook of the planning group and found support through a convergence of the interests of several subject-matter departments within the Authority.

Until 1942, the TVA's policy on land acquisition was formulated separately for each reservoir in a resolution adopted by the Board. Thus, on December 20, 1935, the Board approved a resolution on land purchase for the Guntersville Reservoir (on the main stream of the Tennessee River) which read in part as follows:

> WHEREAS, the Authority desires to start at once the purchases of land in the Guntersville Reservoir area, and
>
> WHEREAS, the Board considers it necessary to purchase a protective belt of land around the reservoir,
>
> BE IT RESOLVED, That the Director of Engineering Service is hereby authorized to approve for purchase the lands to be flooded by the Guntersville Reservoir, and in addition, such lands as are necessary to form a protective belt around the Guntersville Reservoir, extending in general from three hundred (300) feet to one thousand (1,000) feet from the edge of the reservoir, with occasional variations in width for short stretches as are dictated by topography, ownership, or other valid considerations.

The explicit acceptance of the reservoir protective strip as official policy continued, with some variations, until 1941. A belt of publicly owned land 300-1,000 feet in width was also authorized for Chickamauga Reservoir, near Chattanooga, in 1936. A modification was introduced in 1939 with respect to Kentucky Dam, at which time a "reasonable" protective strip was authorized without specifying dimensions. In addition, other criteria for justifiable overpurchase were more specifically delimited: to eliminate peninsulas or tips projecting into the reservoir; to eliminate the danger of erosion and consequent silting by the purchase of steep hillside slopes adjoining the reservoir; to eliminate severance damages by the purchase of inaccessible or worthless farm remnants, islands, and {194} peninsulas; and to avoid the necessity of relocating roads or other facilities when it would be more economical simply to purchase relatively inaccessible areas that would have to be provided with access and other services.

The resolution on Watts Bar Reservoir—behind another main stream dam—adopted February 28, 1940, eliminated the earlier references to the Board's general interest in the purchase of a protective belt around the reservoir (as in the "whereas" clause quoted above), but did authorize the provision of "a reasonable protective strip around the margin of the reservoir." Impractical severance of individual properties and uneconomical provision of access facilities were again cited as justifications for overpurchase, but a new element was introduced in a provision that the right to flood (flowage easements) might be acquired in lieu of fee simple titles "on such lands as are occasionally flooded, where such easement purchases appear practicable and economical." This sequence, while still accepting throughout the principle of a reservoir protective strip, indicated a gradual weakening of Board policy. The weaker policy was transformed into an outright reversal when the gathering strength of those who opposed public ownership in principle became decisive in shaping the policy of the TVA Board.

During this second phase of the land acquisition policy, the several interested departments of the Authority took positions on the issue which reflected their various professional commitments. Those in favor of the policy of providing a protective belt around the reservoirs included:

The professional planners.—After the departure of E. C. M. Richards, and the consequent truncating of the authority of the foresters, the Regional Planning Studies Department took the lead in favoring public ownership of a protective strip as a matter of principle. This group bore much of the brunt of the internal debate, and formulated the principle issues in terms which reflected the broad views of many planning officials. One such formulation was set forth by the leading TVA planner of the first seven years, Earle S. Draper, in 1938:

> In the sense that the building of our reservoirs may have broad relationships and serve as a means to the end of social and economic improvements, the acquisition of reservoir purchase area may relate to regional adjustments as well as to actual water control in such a way as to make rather important demonstrations of the best uses of land. The actual ownership of the protective strip may give indirect benefits in demonstrations relating to farming, forestry, recreation, etc., of great importance to the over-all objectives of the TVA program. {195}

The planning group underwent a series of changes which resulted in a general curtailment of objectives and a "more realistic" outlook, but continued their efforts in support of the protective strip policy until the actual reversal by the Board was consummated.[12]

The foresters.—Although adopting a more limited perspective on land acquisition and land-use policy, the foresters continued to support the idea of public ownership as that was applied in the protective strip policy. The Forestry Relations Department, even under its new leadership, tended to be aligned consistently with the planners in defending the protective strip policy. Moreover, in 1940, Chief Forester Willis M. Baker (who succeeded E. C. M. Richards) stressed the need for the Authority "to recognize more clearly certain fundamental principles and basic objectives relating to land use." Though generally recognized as more moderate in his views than his predecessor, he too was faced with the problem of overcoming pioneer attitudes toward the land. In the course of a discussion of this problem, Baker said:

> In the formulation of a policy based upon local and landowner responsibility, several facts must be kept clearly in mind. The past depletion of lands and natural resources may be attributed largely to our national conception of the rights of the private citizen and to national policies set up to protect these rights even at the expense of public welfare. Our American assumption has always been that private initiative, through self-interest, would find ways of keeping land pro-

ductive. We now discover that this same self-interest, together with lack of concern for the public or the future, has caused the ruin of many lands and the exhaustion of certain resources, resulting in serious social and economic problems. Conditions have changed greatly since the original land policies were adopted. The public is now faced with a situation that demands realization and acceptance of responsibility for remedial action.

The foresters found it necessary to carry on a continuous effort to counteract those forces within the Authority — notably the agriculturists — who maintained the continued cogency of older assumptions concerning public land policy.

The Maps and Surveys Division.—The protective strip policy found additional support from this organization, under the jurisdiction of the Chief Engineer, which had the responsibility of making on-the-spot determinations of the amount of land to be purchased within limits set by the Board of Directors. This group supported the protective strip {196} policy for technical reasons, rather than as a matter of principle. Thus surveying economies might be effected if deviations from the irregular shorelines of the reservoirs were permitted; or it might often be considered advisable to buy more land than that to be covered by water in order to avoid impractical severance, i.e., the purchase of only part of a farm, leaving the most unproductive upland part in private ownership. Since problems of this sort would normally arise, it was considered desirable and economical to permit a reasonable amount of overpurchase. Possibly the social outlook of the chief of this division, who had been associated with A. E. Morgan in the early years of the agency, was a factor beyond purely technical considerations, which influenced the position of the Division on the protective strip policy. The group was placed in a strategic position, since for a long period the Board delegated to the Maps and Surveys Division the authority to approve the purchase of specific tracts of land. It is significant that in 1942, concomitantly with a general abandonment of the protective strip policy, the Division was deprived of its authority to exercise discretion in this respect.

The Reservoir Property Management Department.—Although direct evidence of its importance is lacking, some participants suggest that this department, charged with administering TVA-owned lands, also threw support in favor of the protective strip policy, "in order to protect their reason for being." The abandonment of the protective strip policy not only reduced the extent of TVA-owned lands which would require management, but also carried the threat that the Authority would sell its holdings and thus eliminate much of the department's functions.

The groups so far mentioned constituted the major alignment in favor of the protective strip policy. Other departments, less directly concerned, tended to support the existing policy. But the opposition was powerful, and ultimately prevailed. It included:

The agriculturists.—As TVA operations moved from the tributaries to the main stream of the Tennessee River, and concomitantly with the overturn of personnel which followed the removal of A. E. Morgan in 1938, the TVA agriculturists opened a campaign to effect a reversal of the agency's land-purchase policy. This program included, as will be noted below, a fundamental opposition to the principle of public ownership of land. In pursuing this policy, the agriculturists were reflecting the attitudes and interests of the agricultural institutions of the Valley area.

The Chief Engineer.—After the removal of A. E. Morgan, who had exercised a dominant influence over the construction program, the agriculturists {197} found a welcome ally in the person of a new Chief Engineer, T. B. Parker. Apparently lacking some of the broader interests which characterized the construction leaders of the early period, Parker introduced an economy-minded approach, with general opposition to policies which would add to the cost of the construction program, including the overpurchase of land around the margins of the reservoirs. The organizational weight of the Chief Engineer in exercising discretion and in influencing the development of Board policy, was thrown to the side of the agricultural group. The latter understood the convergences of interests involved and readily admitted that, in effect, an intraorganizational alliance had been consummated. The ascendancy of the new economy-oriented leadership among the construction engineers is a significant symptom of the modification of the character of TVA. It meant that those charged with administering an important phase of the program were no longer dedicated to the use of the technological changes introduced by the agency as instruments for the achievement of tangential social purposes. In a minor way, the Maps and Survey Division—long under the influence of A. E. Morgan's point of view—underwent an experience similar to that of the Forestry Relations Department. Since the Maps and Surveys Division was within the formal jurisdiction of the Chief Engineer, pressure could be put upon it to make an adjustment to new emphases in policy.

The divergences of opinion and interest reflected in the departmental alignment outlined above found expression within a short time following the organizational crisis, congressional investigation, and administrative reorganization of 1937–1938. A debate on land-use and land-purchase policy was initiated by 1939 and continued until 1942, at which time the Board established a new policy, reversing earlier positions and climaxing a steady trend away from the conservationist objectives and methods associated with such agencies as the Forest Service, Soil Conservation Service, Bureau of Agricultural Economics, and National Resources Committee. The nature of the debate, and its sequence, is shown in a series of general statements formulated by the several interested groups within the Authority:

1. In June, 1939, after a year's experience in the Authority, Chief Forester Baker expressed his serious concern with regard to the TVA's

policy on problems of acquisition, use, and management of rural lands. He voiced in a more moderate and restricted vein some of the disagreements which his predecessor had made public. In addition to stating his own views, Baker noted that the Authority was receiving criticism in {198} Washington which expressed "the unfavorable attitude of many representatives of conservation agencies and national associations toward what they regard as the anticonservation activities and practices of the Authority." He suggested that, on the basis of his own experience, there appeared to be some justification for this criticism, requiring the formulation of definite policies and more efficient procedures by the Authority. While proceeding cautiously and apparently with an eye to the strength of the agriculturists within the agency, Baker supported the need for land acquisition in the interests of such purposes as recreation, game refuges, fish hatcheries, tree nurseries, and other public projects. He also obliquely raised the question whether the Authority was not ignoring or overlooking some of its responsebilities in its zeal for dependence upon private initiative and for the protection of local interests.[13]

2. In November, 1939, the Director of the Regional Planning Studies Department (later redesignated simply Regional Studies Department) expressed the belief that the Authority's organization for land-use planning was inadequate. In particular, he observed that the Department of Agricultural Relations did not seem able to execute its responsibilities for agricultural land-use planning. He hinted broadly that the agriculturists did not coöperate with other departments in joint studies of problems involving agricultural planning, and suggested that the official formulation of the duties of the Chief Conservation Engineer over-stressed the agricultural aspect and did not give enough weight to forestry. These matters have already been touched upon above. Significant in this context is that this summary assignment of responsibility for purported defects represented part of the debate over land-use policy, stressing its administrative dimension.

3. In February, 1940, the Chief Conservation Engineer formulated his views on the Authority's land-acquisition policy. Accurately reflecting the views of the TVA agriculturists, he based his proposals for a general land policy on the following propositions:

1. Land can be acquired only "to carry out the purposes of the Act." This should be practically accomplished with maximum economy to the Authority and with the least disruptive effect to the region, taking into consideration fairness to the landowner whose individual property is affected.

2. Where there are elements of doubt as to the amount of land *necessary* for statutory purposes, these doubts should be resolved in favor of leaving the land in private ownership. This is particularly important to the agricultural industry upon which the impact of land removal for reservoir use is necessarily greatest. {199}

The proposed policy definitely cast aside the collateral objective, favored by the planners and others within the Authority, of capturing for the public the increment of value resulting from the construction of reservoirs. In general, the agriculturists have considered the grass-roots policy to be the prime collateral objective of the Authority's activities. In the land-acquisition policy, the grass-roots method was conceived of as demanding minimal purchase at the reservoir margins. Moreover, it was proposed that lands acquired by the Authority for reasons of ultimate economy (farm remnants and inaccessible and isolated areas) should be disposed of by sale as rapidly as possible instead of trying "to find TVA uses for these properties as reasons for continuing to own and operate them." At another occasion during the same period, a representative of the Department of Agricultural Relations put the matter in categorical terms: his department wanted private ownership of land to extend to the water's edge.

4. After considerable discussion, the whole problem of the protective strip was thoroughly reviewed when determining land acquisition policy for the Fort Loudon project in 1941. In May of that year the opposing groups within the Authority made most explicit the varying principles which grounded their judgments and consequently the basis for the enduring disagreements. It was clear that the issue would not be resolved except by some formal step taken by the Board of Directors. Consequently, the General Manager requested the staff to formulate the divergences of opinion for review by the Board.

Provided with this opportunity the chief proponents of the reservoir protective strip and their opponents formulated the principle bases of their respective positions. The protective strip policy was identified as follows:

> a) Land lying below the contour "at the top of the gates," i.e., subject to flooding due to the backing up of water behind the dam, but not the maximum backwater line, would be acquired by outright purchase in fee simple, except where there was clear warrant for the purchase of easements.
>
> b) A protective strip extending back approximately 300 feet from this contour would be provided.
>
> c) Other lands lying between the contour at the top of the gates and the backwater line are acquired by easements except where purchase in fee simple appeared clearly desirable.
>
> d) Fee purchase would be authorized when necessary to avoid impracticable severances, isolation, and road relocation.[14] {200}

The arguments marshaled by the planning group in support of this policy reflected the viewpoint of those who were willing to use government intervention as a tool for the achievement of public welfare objectives incidental to the primary purposes of the agency and subject to its discretion. Four major arguments were set forth by the proponents of the protective strip policy:

i) The protective strip would capture for the public the incremental value created by the new reservoirs—an increment which would otherwise accrue to only a limited number of individuals. This was conceived to be consistent with the policy of recognizing the public interest in another by-product of reservoir construction—electric power. The cost of capturing this increment would be sufficiently low to make it simply incidental to the primary program of the Authority; failure to accept this responsibility would leave the agency open to the justifiable charge of neglecting its rightful obligations to the public. Moreover, the cost of recapturing reservoir sites once a project was completed would normally be prohibitive. Hence a future and foreseeable obstacle to the creation of parks and other projects would be avoided by the reservation of a protective strip.[15]

ii) The protective strip, creating a reservoir area of publicly owned waterfront land, would permit the realization of opportunities for public benefit as they might arise. Though public demand may be slow to crystallize, the technical judgment of a responsible public agency such as TVA should be able to take advantage of foresight in the interests of the general welfare. The agriculturists within the Authority had demanded that land be acquired only for specific purposes for which a current need was evident. The principle of a reserve area to be utilized as occasion arose was formulated in order to counter that demand.

iii) The protective strip would permit the Authority to establish conditions at the reservoir margins which would provide opportunities for private enterprise to operate the shoreline facilities soundly and profitably but without detriment to the public interest. The Authority would at once protect the continued cleanliness and beauty of the reservoir {201} sites and at the same time provide opportunities, through licensing and leasing of public lands, for private investment. It was pointed out that at Guntersville, Alabama, under TVA ownership of the commercial and industrial reservoir frontage, three oil terminals, a sawmill, a sand and gravel company, a grain elevator, a lumber yard, and an automobile transportation terminal had been established within a three-year period.

iv) The protective strip would permit public control of lands at the reservoir margins, with the opportunity of enforcing land-use objectives: the prevention of abuses of land and forest cover and the exclusion of undesirable developments having inadequate standards of safety, sanitation, operation, and appearance. It was stressed that public ownership of the reservoir strip could compensate for the absence of legal control at the local level, and remain the most reliable device for avoiding shacks, billboards, and other structures which would deteriorate the scenic resources of the waterfront lands. Lack of a protective strip in some areas was said to have made possible the development of substandard services by uncontrolled private operators and the over-development of facilities which, if subsequently abandoned, may create backwater slums.

It is apparent that the reasoning of the proponents of the protective strip policy in general accords with the philosophy which underlay the establishment of the Authority itself. On the other hand, the opponents of the protective strip appear to have been indifferent to such considerations. The policy they supported during the 1941 discussion was:

a) Land below the backwater line, but above the *minimum* operating level of the reservoir, would not be purchased outright. Flowage rights would be acquired through the purchase of easements.

b) Lands required for authorized programs and projects of the Authority would be purchased in fee.

c) Lands purchased to avoid excessive costs of highway relocation, isolation, or impracticable severance would be sold concurrently.

The protagonists of this policy based their program upon a set of assumptions which cannot be considered consistent with the broad objectives expressed by President Roosevelt and others who initiated the TVA project. Moreover, it tended strongly to pull the TVA outside of the conservationist movement as that was expressed in the activities and organizations established since the pioneer efforts of such men as Gifford Pinchot.[16] Opponents of the protective strip policy stated their objections. {202}

i) The purchase of a minimum of land "would leave in private control the maximum land suitable for use in the established agricultural economy in an area where it is more urgently needed than in any other region of the nation." In effect, this argument rejected the conception that the introduction of technological changes, such as the impounding of reservoirs, inevitably altering established conditions, would offer opportunities for the achievement of corollary social objectives. No account was taken of accompanying changes that, would be introduced by private owners, such as the establishment of residential subdivisions along the new shorelines.

ii) Minimum land purchase at the reservoir margins would reduce displacement of population and expenditures for readjustment activities. This was one of a series of economy-oriented arguments, which in effect rejected the earlier assumption of responsibility by the Authority for the social consequences of technological change. The burden of caring for population remnants would be shifted to local agencies in many cases inadequately equipped to handle them. While a short-term advantage might be gained, the long-run consequences of avoiding the problem of displacement would ultimately have to be faced in some way.

iii) For this analysis, perhaps the most significant of the arguments proposed in an official way by the agriculturists was that which affirmed that the policy of minimum land acquisition "would provide the opportunity for private enterprise to exercise its initiative in the development of the waterfront free from the stifling effects of public ownership and control and in accord with the public demand." At one stroke, the opposition of the agriculturists to public ownership of land was laid bare. This meant that a change in policy based upon such an assumption would be not merely a matter of expediency but would cut at the fundamental social outlook, hence at the character, of the TVA as a whole.[17]

A REVERSAL OF POLICY

The steady retreat from the policy of providing a protective strip in public ownership around the reservoirs was reflected in an increasing pressure to purchase flowage easements rather than fee simple titles to {203} tracts which would be flooded only occasionally. By late 1941 this practice was already being widely used. It is indicative of the general character and social implications of this trend that even where outright public ownership had been foregone, and only flowage rights procured, the opponents of public ownership pushed their demands so as to eliminate the last elements of public control over land at the reservoir margins. In executing the policy of purchasing flowage easements, the General Manager had authorized the Land Acquisition Department to include a provision in all such contracts permitting public access to the land covered by the easement for purposes of hunting, camping, fishing, emergency landing, and the like. This would permit the reservation to the public of recreational opportunities created by the new reservoir. Under this policy, in case the Authority representatives were unable to acquire flowage easements which included a public access provision, outright purchase would be authorized. The inclusion of the public access provision was considered especially important in such areas as that around Fort Loudon Reservoir (near Knoxville) where a significant movement of population to the reservoir shore line could be expected. If such a movement occurred, and the right of public access from the reservoir to the shore had not been reserved, it was anticipated that soon the general public would be denied use of the lake created by the TVA dam.

The general manager defended the public access provision in the purchase of easements, though he suggested that the elimination of camping as one of the reserved rights might quiet some objections. The opponents of the public access provision within the Authority were speaking for the local landowners; in effect they were ready to relinquish to a relatively few individuals the opportunities for use and enjoyment created as a result of the technological changes introduced by TVA in the area. The opposition to public ownership or control over land use tended to wipe out that concern for the social consequences of its activities which represented an important aspect of the character of the Authority as a whole. The public access provision remained intact, but the significant successes of the agriculturists and their allies within the Authority, in respect to the evolution of the land-purchase program as a whole, seemed to weaken rather than strengthen the general manager's position.

A significant victory for the agriculturists, as was generally recognized within the Authority, came on February 2, 1942, when the Board of Directors adopted a resolution defining a general policy on land {204} acquisition which definitively rejected the policy of acquiring a protective strip. The Board declared that it was now its policy "to limit the purchase of land for reservoir purposes to the minimum appropriate for the particular project. In general, land lying above the zone of reservoir fluctuations" was not to be purchased in fee simple except under certain specified conditions. In effect, the Board accepted the reasoning of those who opposed the protective strip and rejected that of its proponents. The retreat from the early policy was now complete. From the period of the construction of the Norris Dam, nine years earlier, which envisioned and effected widespread public ownership and control of rural lands, the policy had now become one of restricting public ownership to the minimum dictated by specifically authorized projects. The economy orientation of the chief engineer and the agriculturists' representation of local landowner interests seemed to have effectively altered one of the basic aspects of the Authority's program.[18]

The Board of Directors itself is said to have been split over the final decision. It is reported that the resolution of February 2, 1942, was passed in the absence of Chairman Lilienthal and that he was in fact opposed to it. However, owing to another controversy then demanding his attention, the chairman is said to have accepted the decision of the other two members of the Board. It can scarcely be doubted that this was in fact a victory for Harcourt A. Morgan and his group within the Authority, for former Senator James P. Pope—the third member of the Board—seems not to have exercised an independent role at this time.

It is probable that the abandonment of the protective strip policy was aided by the public controversy in 1941 over the construction of Douglas Dam. This project, located on the French Broad River near Dandridge, Tennessee, was hotly opposed by representatives of the local area. The leading antagonist was Senator Kenneth McKellar of Tennessee, and he was supported by Governor Prentice Cooper, as well as representatives of the local farm bureau organizations and President James D. Hoskins of the University of Tennessee.[19] A long-standing argument against the {205} TVA—that it brought more evil by flooding productive agricultural land than the good it created—was again given public attention. This must certainly have strengthened the hand of the TVA agriculturists in their fight for revision of the older policy.

However, it must be emphasized that the TVA agricultural group was opposed not only to the acquisition of clearly valuable agricultural land—though acquisition did not necessarily mean such land would be taken out of production—but to all unflooded land, whatever its adaptability for agricultural purposes. Thus they opposed the acquisition of a large tract of forest land, which was to have been turned over to the

U. S. Forest Service as part of the national forest reserve, on the theory that the Authority had no right to decide that the land should be publicly owned forest. Moreover, having won their point on the protective strip, the agriculturists did not rest content. They continued their opposition to specific acquisitions, as for recreation purposes, and intensified their demand that the Authority reduce its effect upon the agricultural economy to an absolute minimum. The latter objective was maintained even when the cost of providing new means of access to some areas would exceed the cost of acquiring the land, or when families would be left to inadequate school services. In all such conflicts, the agriculturists invoked the authority of the local land-grant colleges whenever possible.

"FANATICISM" AND ADMINISTRATIVE DECISION

The consistent pressure brought to bear upon policy by the agriculturists within the Authority was a source of considerable vexation to other departments. This pressure occurred not only in respect to the reservoir land-purchase policy but on tangential questions as well. We may cite two such issues here, and then attempt to explain, in terms of the foregoing analysis, a frequent comment on the behavior of the agriculturists made by other groups inside TVA.

The close link between the TVA agriculturists and the extension services sometimes resulted in embarrassment to the TVA Land Acquisition Department. The latter was responsible for technically evaluating land to be purchased, and for paying prices that would enable the landowners to relocate without lowering their living standards. The Authority's dual role of, on the one hand, attempting to make its purchases at reasonable cost, and, on the other, of benefiting the farmer as much as possible, might normally be expected to lead to difficulties and disagreements. But these land-purchase activities seemed to find a special source of resistance in the attitude of the agriculturists that other departments within {206} the Authority might not be much concerned about the welfare of farmers in the Valley. Moreover, the extension service in the field, operating wherever land acquisition was carried on, provided the TVA agriculturists with a means of implementing their own views.

The extension service officials in the counties could bolster their influence among the farmers by aiding them when condemnation proceedings were instituted by the Authority. Thus one assistant county agent, whose salary was indirectly paid by the Authority under the joint agricultural program, testified against the Authority at a condemnation hearing. Other extension officials carried on activities in support of the local landowners against the Authority. In addition, the extension officials found ready allies among the TVA agriculturists, which enabled them to press their demands through the TVA staff itself. Presumably, this furthered the prestige of the extension service, since it afforded

them an influence which could be discreetly boasted of to the local farmers.

The TVA agriculturists executed their commitment to the policy
of the extension service with great consistency, and were highly conscious of the need to develop a strategy within the Authority. Thus, in
1941, extension service representatives requested that the TVA provide
an appraisal breakdown to farmers whose property was to be purchased. The agriculturists accepted this policy as their own, and, after
the request was rejected by the Land Acquisition Department,[20] raised
the question among themselves whether it would be wise "to wage a
fight to force release of appraisal breakdown." Moreover, the TVA
Director of Land Acquisition did not hesitate to express his conviction
that joint criticisms by the TVA Department of Agricultural Relations
and the extension service organizations had been transmitted to the
local landowners, and that some local organization had been encouraged whose purpose was to put pressure upon TVA for an increase
in land prices. This situation reinforced and deepened antagonisms
which tended to isolate the agriculturists from other groups within the
Authority.

Another situation which reflected the TVA agriculturists' role of
representing local landowners led to a virtual rupture of relationships,
and considerable bad feeling, between the Authority and another conservation agency, the Bureau of Biological Survey.[21] The Bureau was responsible {207} for the national wild-life conservation program, centering partly around the establishment of wild-fowl and game refuges
in various parts of the country. Early contacts had been initiated by the
TVA foresters under E. C. M. Richards for the operation of a joint program within the Tennessee Valley area and especially on TVA-owned
lands. On July 7, 1938, a Wheeler Migratory Wildfowl Refuge was created by Executive Order of the President. The new refuge comprised an
area of some 41,000 acres within the protective strip purchased by TVA
at Wheeler Reservoir near Decatur, Alabama. The Bureau of Biological
Survey was designated as the agency to administer the refuge, but the
lands remained under the primary jurisdiction of TVA. This meant that
the development of the area would be based largely upon coöperative
agreements and administrative relationships established and maintained by the two agencies.

Contacts between the Bureau and TVA had been continuous since
the establishment of the latter agency in 1933, but from the viewpoint
of the Bureau these relationships had never been very satisfactory. In
addition to some broad criticisms of the nature of the TVA program,[22]
there were two main sources of friction. One of these was the TVA
malaria control program, which involved the use of insecticides on a
large scale and the lowering of water levels to kill mosquito-breeding
vegetation. Both of these practices were considered by the Bureau to be
inimical to wildfowl subsistence needs, and the TVA was requested to
adjust its malaria control program so that the deleterious effect upon

wild fowl would be kept to a minimum. This difficulty, which persisted over several years, appears to have resulted from conflicting technical programs requiring administrative coördination, rather than of any disagreements over principle. However, the second source of continuing friction was somewhat different. The TVA followed a policy of leasing its lands to local landowners for agricultural use, utilizing the extension service organization and voluntary land-use associations to execute the leasing program. This activity was carried on within the Wheeler Refuge area, and as a consequence of what the Bureau considered bad agricultural practices from the standpoint of conservation some rather bitter criticism was launched against the Authority.

Relations between TVA and the Bureau became acute in 1939, largely {208} as a result of the activities and viewpoint of the agricultural group within the Authority. The TVA had made certain commitments to coöperate with the Biological Survey with respect to land practices on Wheeler Refuge. The TVA agriculturists, however, apparently did not feel that such commitments were binding without the approval of the local extension service and county agricultural associations.[23] This had certain significant consequences. In the first place, negotiations carried on between the Bureau and TVA would involve an element of unreality so long as it was assumed by one of the parties that there was still another group, not formally represented, whose agreement would be needed before commitments could be binding. Secondly, the Bureau was being asked, in effect, to submit its technical judgments on wild-life conservation needs and practices to the arbitration of a group of local people whose immediate interests were to be directly affected by the program. The first of these consequences was probably the source of that difficulty within TVA management which left it open to charges of bad faith by the Bureau.[24] The second consequence of the introduction of the local groups as a party jeopardized principles of public control which had been accepted only after a long struggle by those interested in a national wild-life conservation program.

The TVA agriculturists seem to have been bent on pursuing their opposition to the reservation of land for nonagricultural uses. For his part, the Chief of the Bureau of Biological Survey insisted that

> . . . it is not our desire to obstruct the agricultural development of good land or to interfere with the local economy by taking valuable agricultural land out of production. However, certain practices which are now being carried on, such as plowing to the water's edge, destructive and late burning, overgrazing, and disregard of contour cultivation, are certainly not conservation measures in any sense of the word, and it is my belief that the Authority is negligent in allowing such practices to continue.

It is clear that the ultimate soil conservation objectives of the TVA agriculturists were not in principle opposed to those of the Bureau. But the

TVA personnel insisted that progress along that line must be based {209} on the education of local people, through their own institutions; on this view, it would be better to have bad agricultural practices for a while than to remove the land from use by local farmers. It is this difference—not so much of objective as of method—which separated the agriculturists from the conservation tradition which the Bureau represented. Basic assumptions as to method, however, such as the use of public ownership as an instrument of planning and control, become part of the character of movements and organizations.

It was precisely within the sphere of this difference in method that the chief incident which deepened the friction between the two agencies occurred. Recognizing that the TVA had already made some commitments to the local agricultural interests at Wheeler Refuge, it was agreed that representatives of the Bureau would attempt to gain the coöperation of those interests through the local county agents and leaders of the county soil conservation associations. A meeting was held at Athens, Alabama, on May 5, 1939, but, far from permitting the Bureau personnel informal access to the local farm leaders, an unexpectedly formal gathering took place, attended by the assistant director of the TVA Agricultural Relations Department, among others. In a general discussion a TVA representative reportedly spoke disparagingly of the program of the Biological Survey, and suggested to the local farm representatives present that the establishment of the refuge was not necessarily permanent and might be revoked if the members of the local soil conservation association so desired. The public character of this meeting lent a special importance to the remarks of the TVA official, and the Bureau was quick to conclude that members of the TVA staff were deliberately attempting to publicly discredit the Bureau and to undermine its program. The consequences of such activity for amicable relations among public agencies need hardly be labored.

The record of this controversy emphasizes once more the ambivalence of the grass-roots approach. On the one hand, TVA's reliance upon the extension service machinery permits its program to be shaped by local interests and local demands. At the same time, the TVA as a whole is forced into embarrassment and maladjustment with respect to the regular federal organization. Normally any regional agency would be torn between its desire to adapt itself to local interests and at the same time to execute a unified national policy. A series of compromises would be expected. This is likely to occur even in an organization formally centralized, with headquarters in Washington. But TVA has gone farther. In delegating its agricultural program to a group which carries the {210} banner of a local constituency, the Authority comes to reject in principle—not simply as a matter of temporary expediency—those elements of the national program which are opposed by the constituency.

In commenting upon the persistent pressure of the agriculturists to shape the policy of TVA on land purchase and land use, other groups within the Authority have characterized their behavior as fanatical. Not uncommon are exasperated references to the apparent willfulness of the agricultural group, to their easy lapse into "preachy" discussion, to the fact that they often constitute a "minority of one"[25] among the interested departments. These are extreme characterizations, and are not necessarily based upon a sober or rounded appraisal of the role of the Agricultural Relations Department. However, one may ask: what organizational significance can we infer from this impression of "fanatical" behavior?

The behavior of a faction within an organization assumes a quality of strangeness when it appears to be reckless of the basic unity of the whole group.[26] It is the continuity of basic assumption which knits together the several administrative units: it is that continuity which is presumed if methods of formal coördination are meaningful. When, however, a group is divided on matters of fundamental significance, the real locus of decision will tend to move out of the sphere of formal administrative channels. The breakdown of real unity will be reflected in administrative coördination becoming no longer a mechanism for the resolution of technical and transitory maladjustments but rather a vehicle and an instrument in the unacknowledged struggle over a primary prize: the evolving character of the organization as a whole.[27] Thus an examination of the context of the TVA Budget Office move, in 1943, to tighten control over the Agricultural Relations Department[28] reveals that an important motivating factor arose from the doubts felt by the Budget Office personnel that the agriculturists were {211} in fact carrying out a program consistent with the ideals which brought the Authority into being. In general, it was possible to gather a sense of uneasiness with respect to the agriculturists—a feeling that somehow this particular group was not an integrated part of the TVA as a whole and did not completely share its aspirations. This feeling was by no means universal, but sufficiently widespread to offer a basis for comment and reflection.

The characterization of the agriculturists as fanatic doubtless finds its roots in this sense of underlying disunity. For it is this disunity which makes possible action which, from the point of view of the group as a whole, is irresponsible. The quality of responsibility in action arises when the basic commitment of an individual or a faction is to the integrity of the whole; it involves a willingness to compromise, to accommodate special outlooks and goals, and presumes an acceptance of common principles. When, however, there is a sense of isolation, of alienation, action may be shaped by criteria of conflict and disaffection. Moreover, when the cleavage is not fully recognized, other groups may simply feel that "something is wrong" and offer psychological explanations for what appears to be strange and discordant behavior.

With these considerations in mind, it is possible to suggest three factors which, taken together, may offer a suggestive explanation of the

label "fanatic" which has been applied, in some quarters of the Authority, to its agricultural organization.

1. The agriculturists are the only group in the TVA which, in a
unified way, is representative of the local area.[29] The personnel of the
Agricultural Relations Department, and its leadership in particular,
come not only from the area but from a special institution: the agricultural extension services. They share many of the basic attitudes indig-
enous to the area,[30] including hostility to those who come from afar to
effect changes and offer criticism based on alien values. Consequently,
it seems fair to suggest that the agriculturists must look upon the rest
of the Authority as something imposed upon the local area, whose capacity for evil is great and must be limited as much as possible. It is difficult to avoid the conclusion—based upon the arguments offered by
the agriculturists in the land-purchase policy controversy—that it is
only expediency which has restrained them from supporting those[31]
{212} who have expressed the belief that the damage done by the Authority in flooding valuable agricultural lands is greater than the good
accomplished. It is probable that the sense of being representative
gives the agriculturists the feeling that they are playing a historic role
in the interests of their own people and against the outsiders.

2. Evidence has been presented above that a constituency relationship was constructed involving the TVA agriculturists on the one
side and the land-grant college system on the other. On this theory,
much of the behavior of the agriculturists may be viewed as a function
of external commitment. This serves as another factor in explaining
behavior which appears to be irresponsible. Given an external commitment, responsibility becomes divided or even completely referrable
to the loyalties of the subgroup to its constituency. Such terms as "fanatical" are employed when the real basis of irresponsibility is not clearly
understood. The "fanaticism" of American Communists, and their puta-
tive irresponsibility, becomes clear when the commitment of the Com-
munist Party to the regime in the USSR is understood. What is irresponsible from the point of view of American national interests becomes clearly responsible from the point of view of the Soviet Union.
Less dramatically, but by the same logic, the actions of the agriculturists can be fully understood only in relation to their constituency
commitments.

3. "Fanaticism" may also be a label for that type of behavior which
bases itself upon the persistent espousal of a general doctrine or formula, in such a way that issues are not considered upon their merits.
This is a very familiar use of the label in secular contexts. If concrete
problems are simply occasions for the expression of a general point of
view, if the special goal of a particular individual or group is always put
forward without regard to what most participants consider an appropriate occasion, the use of the term "fanatic" is normally in order.
But the characterization may often have little to do with the personality of the individual involved. It is his method which counts, a

method which may be consciously adopted under organizational pressures and in fact inimical to his own psychology. Thus a man with a special doctrinal axe to grind, as, say, that the interests of workers and employers are antithetical, may squeeze doctrinal implications out of each concrete issue, and take his position on the level of doctrine. In the same way, the TVA agriculturists have had certain general goals in mind—the defense and strengthening of the land-grant college system and the limitation of public ownership of land—and they have handled so many concrete issues according to their implications for these goals {213} that others in the Authority have come to think of them as having an unreasonable approach.

Hence we can see the organizational basis for the so-called fanaticism of the agriculturists: alienation, induced by the position of the group relative to others; action as a function of external commitment; and the persistence of intervention on the level of doctrine. At the same time, it is clear that such an explanation could not have been reached apart from the analysis of the concrete meaning of the grass-roots approach.

---

[1] John Dewey, *Human Nature and Conduct* (New York: Modern Library, 1930), pp. 46, 38.

[2] See above, pp. 64 ff.

[3] It would seem that the Personnel Department, in its relation to the A. F. of L. Tennessee Valley Trades and Labor Council, might have a similar opportunity, however. Doubtless much depends on the relative weight of the various constituencies in the area of operation and on the relative loyalties of the several departments to the organization as a whole.

[4] Daniel Bell, in conversation, has made the cognate point that power and the ends of power may not be divorced in a proper sociology. This is an important reminder for all who study the mechanics of organizational interaction.

[5] House Doc. 15, 73d Cong., 1st sess. (1933). Emphasis supplied.

[6] 48 Stat. 58 (May 18, 1933). Emphasis supplied.

[7] L. C. Gray, "Our Land Policy Today," *Land Policy Review*, Vol. 1, No. 1 (May-June, 1938).

[8] U. S. National Resources Planning Board, *Public Land Acquisition in a National Land-Use Program, Part I: Rural Lands* (Washington, D.C.: U. S. Government Printing Office, 1940), p. 2.

[9] *Ibid.,* p. 3.

[10] We have noted above, pp. 58-59, that such conceptions as "water control on the land" serve to tie general welfare objectives (e.g., soil conservation) to the constitutional pegs of navigation and flood control. The need for oblique justifications of developmental programs has been strong because of the relatively weak delegations in the statute. This tendency to dissemble was, apparently, one of A. E. Morgan's objections to the administrative methods of the majority bloc. Moreover, in respect to Richards' strictures, it should be recalled that during the early days of the Authority some controversy arose between those who conceived of the project primarily as an electric power business and those who placed its developmental functions in the forefront. To the extent that developmental functions were chargeable to power operations, this would increase the cost of production and thus endanger the "yardstick" objectives of electric power operations.

[11] See also E. C. M. Richards, "The Future of TVA Forestry," *Journal of Forestry,* XXI (July, 1938), 643-652.

[12] These changes included some overturn in personnel, a change in title of the Regional Planning Studies Department by omitting the word "Planning," an abandonment of the concept of over-all planning for the Tennessee Valley analogous to that carried on for the nation by the National Resources Planning Board, and the loss of authority to conduct planning studies in rural land use. The course of the regional planning group inside TVA is not unlike that which led to the curtailment of the functions of the NEPB and its eventual elimination.

[13] See the report below of the land-use controversy involving the Biological Survey, pp. 206 ff.

[14] These criteria had been earlier accepted for purchase at Watts Bar Dam, with the exception of the width of the protective strip which in that case— over the objection of the planners—had been set at 50 feet. In formulating their general policy, those in favor of the protective strip were asking for one more generous than that accepted for Watts Bar.

[15] Various examples of increases in land values on the chain of lakes created by the reservoirs were cited, including the following: at Guntersville, Alabama, an increase for some tracts of as much as 5,000 per cent since the creation of Guntersville Lake; near Caryville, Tennessee, the donation of 150 acres of land for park purposes was thwarted when a large portion of the area was withdrawn for a residential subdivision; at another park site, the cost of supplementing TVA purchases by two small additional tracts of land was materially greater than the cost of similar tracts within the TVA strip acquired before the filling of the reservoir.

[16] In 1943, the general manager explicitly recognized that TVA was not within the traditional conservation movement. As formulated by the agriculturists this meant that TVA was opposed to the trend of setting land

aside under public ownership; or, as sometimes alternatively expressed, that TVA was interested in land in relation to people, not in the land for its own sake.

[17] The question of the practical desirability of such a course, or of the agriculturists' view, is not here in question. Important is that the controversy involved a challenge to the older ideals accepted by the Authority, and consequently represented pressure to effect a change in the basic character and meaning of the organization. This is clarified even further by the agriculturists' assertion, often repeated, that "there is really no such thing as sub-marginal land."

[18] The agriculturists themselves were not overly concerned over program economies as such; they were in favor of paying high costs for road relocation or severance damages if that would permit land to remain in private ownership.

[19] The public opposition of President Hoskins was a blow to the TVA, which had long coöperated with the University of Tennessee, granting it administrative control over personnel subsidized by TVA funds. In the Douglas Dam controversy, the TVA's commitments to the defense program defined it as a representative of the federal government in the Valley. To the extent that the Authority had to identify itself with the federal program, it had to bear the brunt of opposition to that program. TVA would have liked to have its cake and eat it too: contribute mightily to the federal defense program and still retain the loyalties of those local institutions and people who would be adversely affected.

[20] On the grounds that an appraisal breakdown furnished to property owners would undermine the purchase program, since it would provide landowners with additional grounds for contesting the Authority's prices; i.e., appraisals of parts of a property, rather than the property as a whole, would become subject to debate.

[21] Transferred July 1, 1939, from the Department of Agriculture to the Department of the Interior, later constituting part of Fish and Wildlife Service. The chief of the Bureau, Dr. Ira N. Gabrielson, became chief of the Service.

[22] In 1934, the then chief of the Bureau criticized the TVA for failing to put its wild-life program on an economic basis, saying that the TVA seemed to consider that program "as important a unit in their plans as a caged canary bird hanging in the dining room." This criticism seems to parallel the strictures voiced by TVA's first chief forester.

[23] In effect, these county associations were creatures of the extension service, run by the county agent's office. See pp. 231 ff.

[24] These were strongly put, and repeatedly. On one occasion, in 1939, the chief of the Bureau commented that "agreements, written or verbal, have been of little consequence to the TVA." The seriousness of the breakdown

of amicable relationships between TVA and the Bureau was attested to by Chief Forester Baker in 1940. He suggested that failure on the part of the Authority to resolve the critical situation might well lead to action on the part of the Bureau—by that time known as Fish and Wildlife Service—which might bring about a Congressional investigation, or some "similar action equally disastrous." The Chief Forester himself was apparently in basic sympathy with the Service, but was helpless because of his relatively weak organizational position within TVA.

[25] Which need not mean defeat, since decisions are of course not made by taking a vote of staff members or sections.

[26] This whole discussion of "fanaticism" is informed by Max Weber's distinction between *Gesinnungsethik* ("ethic of ultimate ends") and *Verantwortungsethik* ("ethic of responsibility"). The peculiar role of the agriculturists becomes quite understandable when it is seen as ordered by ends which are ultimate or absolute with respect to the organizational system in which they are participants. See H. H. Gerth and C. Wright Mills, *From Max Weber* (New York: Oxford University Press, 1946), p. 120 f.

[27] This principle is more obvious in, but by no means restricted to, formally democratic organizations, as when, in periods of dispute, the meetings of an executive committee may serve only to preserve the fiction of unity whereas the real decisions are being made on a factional basis, *in camera*.

[28] See above, p. 115.

[29] See above, p. 111 f.

[30] "As between pellagra and tourists," one of the leading agriculturists is reported to have said, "I'll take pellagra."

[31] Such as Andrew J. May of Kentucky, formerly chairman of the House Military Affairs Committee.

Part Three

AMBIGUITIES OF PARTICIPATION

CHAPTER VII

THE VOLUNTARY ASSOCIATION AT THE END POINT OF ADMINISTRATION

In our state, we feel that the extension service must control the system of community organizations, county associations, and neighborhood leaders. Other agencies must come through the extension service. After all, it's our baby.

A SUPERVISOR ON THE JOINT TVA–EXTENSION
SERVICE PROGRAM (1943)

THE CONSTRUCTION of an administrative constituency, whereby the dominant agricultural leadership in the Tennessee Valley area was afforded a place within the policy-determining structure of the TVA, is an example of the process of *informal* coöptation. The preceding chapters have traced the origin and structural details of the constituency relationship and have sought out its unanticipated consequences for the role and character of the Authority. The explication of these consequences functions at once as evidence for, and as a key to the meaning of, the coöptative mechanism. The unacknowledged absorption of nucleuses of power into the administrative structure of an organization makes possible the elimination or appeasement of potential sources of opposition. At the same time, as the price of accommodation, the organization commits itself to avenues of activity and lines of policy enforced by the character of the coöpted elements. Moreover, though coöptation may occur with respect to only a fraction of the organization, there will be pressure for the organization as a whole to adapt itself to the needs of the informal relationship. Viewed thus broadly, the process of informal coöptation represents a mechanism of comprehensive adjustment, permitting a formal organization to enhance its chances for survival by accommodating itself to existing centers of interest and power within its area of operation.

This analysis directs attention to the unavowed meaning of the official grass-roots policy. One of the major tenets of that policy is the injunction that the program of a regional agency be channeled through the existing institutions of the area of operation, and that a positive policy of strengthening those institutions be maintained. In formal terms, this is taken to mean that the institutional resources of the local area will be fully utilized, and a democratic partnership with the people's institutions effected. The informal organizational consequences of that injunction depend on the relative strength and the respective problems {218} of the organizations involved, and are not readily apparent before analysis. It is that analysis which the preceding chapters have attempted to provide.

But there is another phase of the grass-roots policy which deserves similar attention. The concept of a regional partnership includes the idea that ordinary citizens will be drawn into the administration of the regional program through membership in voluntary associations.[1] "Wherever the execution of the over-all program reaches out into a local community, it is considered desirable to organize those citizens most closely affected into an association which will participate in the administration of the program. There is thus envisioned a mushrooming of citizens' organizations at the end point of administration, permitting participation at the grass roots in the application of a general policy to varying local conditions. The TVA has raised this procedure to an administrative principle, treating it as a fundamental aspect of what it considers to be a unique grass-roots approach.

The policy of encouraging citizen participation through the creation of voluntary associations should be clearly differentiated, in terms of its implications for democratic administration, from that approach which assumes the legitimacy of certain people's institutions, through which action at the grass roots must be channeled. The use of voluntary associations does not of itself determine whether the existing institutions of an area are to be utilized. Indeed, the construction of voluntary associations *ad hoc* may serve as a substitute for the channeling of a program through established organizations. Associations created through the direct intervention of a federal agency, to implement the administration of its own program, may result in the establishment of new channels to the population, by-passing an existing leadership or established institutions. This possibility is at hand whenever an appeal is made to the individual citizen rather than to a previously instituted local leadership. Therefore it is not always clear that these two separate tenets of the grass-roots approach reinforce or even necessarily complement each other. In creating the electric power Coöperatives, the Authority reached out directly to organize the local citizens for participation in the execution of the program of the regional agency. This represented the creation of new centers of organizational strength without the mediation of any previously existing institutions.[2] In the agricultural program, the Authority has made working with and {219} through the extension service organizations its primary grass-roots policy. Thus even within TVA experience, the use of voluntary associations may or may not he connected with the policy of siphoning the program through established institutions. In general, it appears that specific programmatic and administrative imperatives, rather than general considerations of democratic policy, determine the form of intervention at the grass roots.

It follows that the significance of two tenets of the grass-roots theory—working through existing institutions and using voluntary associations—must differ considerably. We have already interpreted the one as informal coöptation, the adjustment to concrete forces within the Valley. In approaching the other, we shall utilize the concept of formal coöptation, in which is uppermost the need to share the responsibilities for or the administrative burdens of power, rather than power itself.[3]

FORMAL COÖPTATION AND AGRICULTURAL DEMOCRACY

The use of voluntary associations is not new, and is far from unique or peculiar to the program or administration of the Tennessee Valley Authority.[4] Indeed, it is useful to think of the coöptation of citizens into an administrative apparatus as a general response made by governments to what has been called "the fundamental democratization" of society.[5] The rise of the mass man,[6] or at least the increasing need for governments to take into account and attempt to manipulate the sentiments of the common man, has resulted in the development of new methods of control. These new methods center about attempts to organize the mass, to change an undifferentiated and unreliable citizenry into a structured, readily accessible public. Accessibility for administrative {220} purposes seems to lead rather easily to control for the same or broader purposes. Consequently, there seems to be a continuum between the voluntary associations set up by the democratic (mass) state—such as committees of farmers to boost or control agricultural production—and the citizens' associations of the totalitarian (mass) state. Indeed, the devices of corporatism emerge as relatively effective responses to the need to deal with the mass, and in time of war the administrative techniques of avowedly democratic countries and avowedly totalitarian countries tend to converge.

Democracy in administration rests upon the idea of broadening participation. Let the citizen take a hand in the working of his government, give him a chance to help administer the programs of the positive state. At its extreme, this concept of democracy comes to be applied to such structures as conscript armies, which are thought to be democratic if they include all classes of the population on an equal basis. If analysis and appraisal is to be significant, however, it is necessary to inquire into the concrete meaning of such an unanalyzed abstraction as "participation." In doing so, we shall have to distinguish between substantive participation, involving an actual role in the determination of policy, and mere administrative involvement. In the conscript army, we have a broadening involvement of citizens, with a concomitant abdication of power. The same may be said of the Japanese *tonari gumi*, neighborhood associations which helped to administer rationing and other wartime programs. Such organizations, which have had their counterparts in many parts of the world, involve the local citizens, but primarily for the convenience of the administration. It is easy enough for administrative imperatives which call for decentralization to be given a halo; that becomes especially useful in countries which prize the symbols of democracy. But a critical analysis cannot overlook that pattern which simply transforms an unorganized citizenry into a reliable instrument for the achievement of administrative goals, and calls it "democracy."[7]

The tendency for participation to become equivalent to involvement has a strong rationale. In many cases, perhaps in most, the initiation of local citizens' associations comes from the top, and is tied to the pressing problem of administering a program. The need for uniformity {221} in structure; for a channel through which directives, information, and reports will be readily disseminated; for the stimulation of a normally apathetic clientele; and for the swift dispatch of accumulated tasks—these and other imperatives are met with reasonable adequacy when involvement alone has been achieved. Some additional impetus, not provided for in the usual responsibilities of the administrative agency, is normally required if the process is to be pushed beyond the level of involvement. Indeed, it is doubtful that much can be achieved beyond that level. Such associations, voluntary or compulsory,[8] are commonly established *ad hoc*, sponsored by some particular agency.[9] That agency is charged with a set of program responsibilities. These cannot be readily changed, nor can they be effectively delegated. As an administrative organization, the agency cannot abandon the necessity for unity of command and continuity of policy—not only over time but down the hierarchy as well. What, therefore, can be the role of the coöpted local association or committee? It cannot become an effective part of the major policy-determining structure of the agency.[10] In practice only a limited sphere of decision is permitted, involving some adaptation of general directives to local conditions, and within that circumscribed sphere the responsible (usually paid) officials of the agency will play an effective part.

With these general considerations in mind, it may be well to mention at least one phase of the historical context within which the TVA's use of voluntary associations has developed. Especially in the field of agricultural administration, the TVA's methods have paralleled an emerging trend in the administration of the federal government. This is not often recognized within the Authority, but there can be little doubt that the United States Department of Agriculture has gone much farther in developing both the theory and the practice of citizen participation than has the TVA. The emergence of this trend accompanied the construction of a vast apparatus to administer an action program reaching virtually every farmer in the nation. {222}

One formulation of the idea of "agricultural democracy" was undertaken in 1940 by M. L. Wilson, Director of Extension Work of the Department of Agriculture.[11] Wilson noted the movement toward a greater group interest on the part of farmers, and the pressure for equality through government intervention, culminating in the enactment of the Agricultural Adjustment Act of 1933 and subsequent New

Deal agricultural legislation. The administration of the new government programs was based on the ideal of coöperation and voluntary participation, leading to a set of procedures which, in Wilson's view, can be thought of as the general principles of agricultural democracy:

1. Decentralized administration in varying degrees through community, county, and state farmer committees, elected by coöperating farmers or appointed by the Secretary of Agriculture.

2. The use of referendums in determining certain administrative policies, especially those having to do with quotas, penalties, and marketing agreements.

3. The use of group discussion and other adult education techniques as a means of promoting understanding of the problems and procedures involved in administration of the various programs and referendums.

4. Coöperative planning in program formulation and localization of programs.[12]

This program emphasizes the importance of participation within the democratic pattern of culture. Moreover, in theory, participation includes both policy-forming and administrative functions.

The technique of coöpting local citizens through voluntary associations and as individuals into the administration of various agricultural programs was widely developed during the nineteen-thirties. In 1940, it was reported that over 890,000 citizens were helping to plan and operate nine rural action programs:[13] community, county, and state committees of the Agricultural Adjustment Administration, operating through over 3,000 county agricultural associations; land-use planning committees organized through the Bureau of Agricultural Economics; farmer associations aiding in the administration of Farm Credit Administration loans; rehabilitation and tenant-purchase committees {223} organized by the Farm Security Administration; local district advisory boards for the Grazing Service; Coöperatives dealing with the Rural Electrification Administration; governing boards of soil conservation districts serviced by Soil Conservation Service; these together form an administrative pattern of which the TVA ventures along this line were only a part. Mr. Ball's summary of participating citizens is reprinted as table 9.

The trend toward coöptation of farmers in the administration of a national agricultural program reached a high point with the organization of the county land-use planning program in 1938. The idea of democratic planning with farmer participation was given considerable attention, and an attempt was made to construct a hierarchy of repre-

TABLE 9

SUMMARY OF ASSISTING CITIZENS IN U. S. AGRICULTURAL PROGRAMS 1939

Name of citizen group	Members
AAA: local committees	135,591
County land-use committees	72,000
Extension Service: volunteer program leaders	586,600
FCA: association directors and committees	36,574
FSA: local committees	26,753
Grazing Service: district advisory boards	547
REA: association directors	4,900
Soil conservation district supervisors	855
Tennessee Valley committees and test-demonstrators	29,035
	892,855

SOURCE, Table 9: Carleton R. Ball, "Citizens Help Plan and Operate Action Programs," *Land Policy Review* (March-April, 1940), p. 26. In 1942, just prior to the withdrawal of federal support from the program, membership in state, county, and community land-use planning committees reached 125,000, according to the *Report of the Chief of the BAE*, 1942.

sentative committees which would embody the democratic ideal.[14] At the same time, the achievement of a primary administrative objective was envisioned. The land-use planning organization program received its impetus from a conference of representatives of the Department of Agriculture and the Association of Land-Grant Colleges and Universities held at Mt. Weather, Virginia, in July, 1938. The Mt. Weather {224} Agreement[15] recommended a system of coördinated land-use planning to overcome the confusion created by the existence of a large number of points of contact between governmental agencies and local farmers. By providing that local officials of the national agricultural agencies would be represented on the farmer committees, it was felt that a single point of contact would be established. It is possible that the land-use planning system would have been established without this impetus from a pressing administrative imperative, but it is clear that the latter was the occasion for the new organization. The problems of the officialdom were primary, and logically so, for their responsibilities had to do with the efficient execution of statutory programs—not the creation of new culture patterns. The latter might, time and resources permitting, have become an effective collateral objective, but it would be idle to suppose that the requirements of administration would not assume priority within the system.

The coöptative construction of systems of voluntary associations fulfills important administrative needs. These are general, and include:

1. The achievement of ready accessibility, which requires the establishment of routine and reliable channels through which information, aid, and requests may be brought to segments of the population. The committee device permits the assembling of leading elements on a regular basis, so that top levels of administration may have reason to

anticipate that quota assignments will be fulfilled; and the local organization provides an administrative focus in terms of which the various line divisions may be coördinated in the field.

2. As the program increases in intensity it becomes necessary for the lower end of administration to be some sort of group rather than the individual citizen. A group-oriented local official may reach a far larger number of people by working through community and county organizations than by attempting to approach his constituency as individuals. Thus the voluntary association permits the official to make use of untapped administrative resources.

3. Administration may be decentralized so that the execution of a broad policy is adapted to local conditions by utilizing the special knowledge of local citizens; it is not normally anticipated, however, that the policy itself will be placed in jeopardy. {225}

4. The sharing of responsibility, so that local citizens, through the voluntary associations or committees, may become identified with and committed to the program—and, ideally, to the apparatus—of the operating agency.

These needs define the relevance of the voluntary-association device to the organizational problems of those who make use of it. It is only as fulfilling such needs as these that the continuity—in both structure and function—of this type of coöptation under democratic and totalitarian sponsorship can be understood.

From the above it is not surprising that criticisms of the county planning program have stressed deviations from the democratic ideal, particularly in lack of representativeness and the tendency for established organizations such as the Farm Bureau to take control of the local committees.[16] Insofar, however, as this represents criticism of a program developing toward complete fulfillment of the ideal, it is not basic. More significant for this analysis are such criticisms as the following:

> . . . it is the central thesis of this paper that county planning did not succeed because no desire to solve community and county problems was created in the population of the area in which the county planning program was to function.... Most administrators of county planning conceive of rural planning as another administrative problem, as a procedure.[17]

The normal pattern—perhaps inevitable because of the rapid creation of a nationally ramified system of committees—established an organization set down from above, oriented toward the administration of the national program. As a consequence, the problems of the local official *qua* official assumed priority. "One needed only to talk with representatives of the several agencies engaged in trying to 'enforce' the county planning system to recognize how ubiquitous this condition [of apathy] was."[18] To the extent that the problems of the officialdom are sufficiently pressing to stamp the character of the organization, it may be expected that involvement rather than meaningful participation will prevail. This same point is made in another way by John D. Lewis, in tracing one of the bases for the lack of complete representativeness. {226}

> The pressure to "get things done" has tended to encourage appointment rather than election. The Division in Washington naturally expects its field agents to report results that will justify the high hope with which the program was launched, and the state office in turn pushes the county agents for progress reports with which to appease Washington administrators. Democratic procedure is notoriously slow procedure. Consequently the first thought of an overworked county agent, unless he is genuinely impressed by the importance of finding a truly democratic committee, will be to find a group of industrious and coöperative farmers who can be depended upon to work together harmoniously. With the best of intentions and with no thought of deliberately stacking the committee, he may set up a committee of "outstanding leaders" who have a sincere desire to act in the interest of the whole county, but who have only a second-hand knowledge and indirect concern about the problems of less successful farmers in the county.[19]

In effect, those responsible for organizing the system of committees or associations are under pressure to shape their actions according to exigencies of the moment, and those exigencies have to do primarily with the needs of administration. As the needs of administration become dominant, the tendency for democratic participation to be reduced to mere involvement may be expected to increase. At the extreme, the democratic element drops out and the coöptative character of the organizational devices employed becomes identified with their entire meaning.[20]

EVOLUTION OF TVA INTEREST IN COÖPERATIVES

The pattern of formal coöptation of local citizens has been, as we have seen, rather widely developed in connection with major programs of the federal government. Recognition of this permits us to examine similar developments undertaken by TVA without ascribing to them a uniqueness which they do not possess. By the same token, an analysis of the general pattern may be expected to retain its significance when applied to the TVA ventures.

The existence of formal coöptation cannot be inferred directly from the bare form of any given organizational device, for it is its function in fulfilling organizational needs which is decisive. Thus one cannot conclude that merely because the TVA has been interested in the development of local coöperative enterprises that these have functioned, or were intended to function, as a means of mustering popular support {227} for the agency. Some TVA officials have indeed expressed the opinion that the voluntary associations developed in connection with Authority programs do operate to create a mass base for the agency. But the available evidence does not seem to warrant any general conclusion that there has been organized a manipulable source of public support. However, insofar as devices have been created which permit the sharing of responsibility for specific aspects of the program, as well as the solution of administrative problems, it does make sense to interpret the use of voluntary associations by the TVA as formal coöptation.

The TVA's interest in the organization of local coöperative associations began with an emphasis on the development of coöperatives as such; coöperation as a form of economic enterprise was furthered in the interest of the general welfare of the region. However, as the programmatic orientations of the Authority came to be more clearly defined—and as certain earlier emphases were stripped away—the interest in local voluntary associations increasingly stressed the implementation of specific subject-matter obligations. In this way the voluntary association device came to be closely tied to the substantive responsibilities and administrative needs of the agency. Concern for the creation and strengthening of coöperatives was not abandoned. But this concern was shaped, and to some extent inhibited, by the problems and emphases of the major programs in connection with which the voluntary associations were organized. The history and rationale of this development may now be reviewed, with particular reference to organizational issues.

Early interest in the furtherance of coöperatives as part of the general-welfare objectives of the Authority seems to have come primarily from Chairman Arthur E. Morgan.[21] He had long been interested in the development of local self-help programs and was quick to turn the attention of the Authority in this direction. The opportunity for initiating such a program came from a rather devious source, and resulted in some confusion, both administrative and legal. Late in 1933, negotiations were initiated between the Authority and the Federal Emergency Belief Administration to have the latter agency allocate a portion of its funds to a Tennessee Valley project for the organization and financing of coöperatives. Presumably it was felt that the objectives of both agencies would be served, if the funds were utilized to help curtail immediate demands for relief, and, at the same time, {228} to provide a demonstration of methods which could help to raise the level of living in the area.

As a consequence of these negotiations, a corporation was incorporated on January 25, 1934, under the laws of the State of Tennessee, as the Tennessee Valley Associated Coöperatives, Inc., to receive and administer the FERA grant. The incorporators were David E. Lilienthal, Harcourt A. Morgan, and Arthur E. Morgan, then directors of the TVA, acting in their private capacity as individuals. By March of that year an outright grant of $300,000 had been received by TVAC from FERA, and in due course an administrator was appointed. Until 1936, the administration of the projects organized through TVAC represented the sum of TVA activities in this field. No additional funds were authorized, but during this period TVA assumed responsibility for the administrative expenses of the TVAC.

The work of TVAC consisted of financial assistance and advisory service extended to coöperatives organized in certain depressed areas of the Valley, especially in western North Carolina and eastern Tennessee. Two groups of coöperatives received primary attention. One, located primarily in the mountainous areas of western North Carolina, consisted of coöperatives organized to market fruit and vegetable produce and operate a number of small canning plants. A regional association to serve these coöperatives was established, known as Land o' the Sky Mutual Association, through which the Authority personnel aided the local groups. The other major project resulted from a study conducted by the Women's Bureau of the United States Department of Labor, which had investigated conditions under which a large number of persons living in the Southern Appalachian Mountain Region were producing and marketing handicraft articles as an important source of family income. It was believed that the low cash income of these people might be raised through a coöperative marketing project, and so a regional organization was formed known as Southern Highlanders, Inc., through which financial and advisory service was extended. In addition, a number of miscellaneous coöperatives received loans from TVAC, and some experiments in new processing ventures were undertaken.

In 1935, a Coöperative Research and Experiment Division was established within the Authority, primarily to administer the work of TVAC. The authority for the work of the Division was based on Section 22 of the TVA Act of 1933, which permits "studies and experiments . . . for the general purpose of fostering an orderly and proper physical, economic and social development" of the area.[22] This meant that the work, in order to maintain its legal foundation, had to be framed in terms of research and demonstration. The Authority did not have power to accept the value of coöperatives as given and then conduct aggressive promotional activities. The weakness of this delegation of authority in the organic Act underlies much of the ambiguity of TVA developmental programs.

After some disquiet as to the tenability of the relation between TVA and TVAC, and perhaps as a minor by-product of the split within the TVA Board of Directors, it was decided to separate the two organizations. In March of 1936, two of the three members of the TVA Board, Harcourt A. Morgan and David E. Lilienthal, resigned as directors of TVAC. They were replaced by Ray Crow, head of the Alabama Relief Administration, and Clarence Pickett, advisor to the Resettlement Administration. Arthur E. Morgan retained his place on the TVAC Board and continued in this capacity after leaving the Authority. In December, 1937, administrative relations between the Authority and the TVAC were terminated, although some payments under contracts with Southern Highlanders and Land o' the Sky continued into 1938.

The separation from Tennessee Valley Associated Coöperatives marked the end of the period wherein self-help coöperatives were promoted or aided for themselves, apart from specific program objectives of the Authority. Some doubt has been expressed within the agency as to the value of the work which was carried on, and especially as to whether the limited range of activities made possible any judgment as to the appropriateness of coöperative enterprise in the Valley. Of course, such a question can be seriously posed only if it is considered to have been the real objective of the program to carry on "studies, experiments, and demonstrations" as provided for in the Act. In fact, the organizers of the program were probably quite convinced of the feasibility of coöperative enterprises as such and were seizing the opportunity offered by the relief grant to undertake their development in the Valley. Indeed, as one TVA staff member put it, "the sizeable expenditures incurred by the Authority do not seem justified by actual accomplishments in terms of experiments of feasibility; they appear more understandable if regarded as direct action to improve the general welfare in the absence of adequate methods (at the time) for organizing local initiative to do the job." {230}

Reorientation of the work of the Coöperative Research and Experiment Division necessarily followed, with emphasis placed first on a study of all coöperatives in the Valley and then on linking actual aid to coöperatives with the broader programs of the Authority. At the same time, the division was placed under the wing of the agriculturists, eventually being renamed Rural Coöperative Research Division. A series of studies—of farmer marketing-purchasing coöperatives, soil conservation associations, and rural electric coöperatives—was undertaken in conjunction with the land-grant colleges, but by 1943 the major work of the division centered around the extension of advisory services to voluntary associations organized in connection with other TVA projects. The latter included primarily (1) soil conservation associations handling Authority fertilizer, which might also develop into marketing and purchasing coöperatives; (2) land-use associations participating in the rental of Authority-owned lands; (3) rural electric power associations acting as distributors for TVA power; and (4) associations using processes and equipment developed in the course of research sponsored by the TVA Commerce Department. The division was small, consisting in 1943 of three specialists in coöperatives and one accountant. At that time, the yearly budget of the division amounted to approximately $25,000, and about $250,000 had been expended by TVA on work with coöperatives since 1934.

This evolution, exhibiting a marked change in direction and emphasis, is important for the grass-roots approach. Linking the development of voluntary associations to substantive programs, such as the distribution of fertilizer and electric power, affords an opportunity for the participation of local citizens in the administration of those programs. Random organization of coöperatives as ends in themselves does not lend itself to that type of administrative partnership or involvement.

COÖPTATION IN THE FERTILIZER PROGRAM

The involvement of local citizens in the fertilizer test-demonstration program[23] takes place largely through the county soil improvement associations.[24] The organization of these groups is the responsibility of the agricultural extension services and is not undertaken by TVA personnel. The associations select test-demonstration farmers within the county and distribute TVA phosphates. Freight charges on the latter must be paid by the local farmers, but this is always accomplished {231} through a county association. Carlot shipments are the rule, for TVA does not consign fertilizer directly to individual farmers. Thus, if only in order to fulfill this formal requirement, there is a county soil-improvement association wherever the test-demonstration program operates—119 of them in the Valley area.

In terms of the grass-roots doctrine, the proliferation of county associations through the Valley, at the end points of the fertilizer distribution process, represents a mechanism whereby every participant in the program may have a share in making decisions which affect him. However, in pressing inquiry as to whether the concept of formal coöptation applies in this situation, it is necessary to look toward the functions which these organizations serve in fulfilling the needs of those who organized them. Based upon information provided by participants in TVA and in the extension service, the following points may be adduced as suggesting the coöptative character of the soil improvement association:

1. The county associations solved an administrative problem of the extension service by relieving extension workers of official responsibility for the handling of money. It is in the interest of the colleges to preserve the status of county agent personnel as educational rather than administrative, and in any case the extension service does not have and may not desire authority to undertake the execution of non-educational functions. The ideal case is that in which the county agent merely advises local groups of farmers. However, the agent in the field is involved in a continuous dilemma. He cannot very well accept responsibility for pushing a program and at the same time ignore the necessity for implementing it on his own initiative where other hands are lacking; nor can he accept responsibility and ignore the need for measures of control over the administrative process even if carried on by others. This difficulty becomes especially evident when an extension-service worker is actually paid to carry on a definite program with

concomitant pressure for results, as is true of the assistant county agents who receive their salaries out of TVA funds. Consequently, there is a strong tendency for rather tight control over the program to be maintained, utilizing the county associations as mechanisms for a formal shifting of responsibility. The role of the extension worker in such circumstances is analogous to that of the executive secretary of a largely paper organization, who often is in such complete control and so indispensable to its functioning that he is the organization.

2. Although often and, especially, early in the program, the extension agents themselves chose the participants in the test-demonstration work, {232} it became apparent that it would be useful to have an organization through which the selections could be made. Since many farmers thought of the program as essentially the distribution of "free fertilizer," opportunities for complaints that the county agent's office was "playing favorites" would certainly arise. Thus if choices were formally made through some sort of farmer committee, complainants could be referred to the committee and thus responsibility would be shunted away from the county agent. At the same time, through his personal influence and as a result of a general apathy, the agent would be able to see to it that the farmers he wanted would be included in the program. In many cases, the action of the association on selections represented simply a formal confirmation of the choices of the agent. This does not mean that the county agent always desired to make his own choices. On the contrary, the association device would be welcome to him precisely because it might operate to relieve him from the pressure of influential individuals. However, though he might prefer to leave selections solely to the farmers as a group, responsibility for results would necessarily force him to intervene in order to see that a sufficient number of reliable elements were included to assure some success to the program in the county. Here we have an example of the principle (see below, p. 261) that formal coöptation requires the exercise of informal control over the coöpted elements.

3. Indicative of the character of the county associations as administrative arms of the extension service is that the organization of such associations is provided for in project agreements between TVA and the agricultural colleges, but the soil improvement association itself is never a party to such an agreement. The project agreements lay down policy governing the test-demonstration program within the state, and it would seem that if the local groups of citizens are actually to participate in the determination of policy it would be necessary for that participation to reach above the county level. But it is not readily conceivable that the extension service would accept responsibility for the execution of the test-demonstration program and at the same time agree to significant participation by a major third party. In general, references to the county associations in administrative documents support the idea that they are thought of as tools utilized by a particular agency for the implementation of its program.

217

4. There is a general and readily observable identification of the machinery of the county agent system with the local associations. The headquarters of the local association is normally the extension service {233} office in the county courthouse, its records are in the county agent's files, its gatherings are under his auspices. In 1942, after eight years of operation of the test-demonstration program, report forms filled out in connection with a survey of the progress of the soil improvement associations were in most eases signed by the assistant county agents. The official *Test-Demonstration Information Book*, published in 1943, informs us that "early capital for many associations was acquired from a handling charge of a few cents on each bag of phosphate issued to demonstrators. Such funds are widely used for educational purposes such as the purchase of cameras, projectors, film, and slides *for the county agent's office*." (Emphasis supplied.)

5. Perhaps the crucial test of control over coöpted citizens' groups is this: is access to the association by outside elements channeled through officials of the coöpting agency? If it is understood that accessibility is a primary objective of the coöptative mechanism, then it may be anticipated that such organizations will be looked upon as the "property" of their initiators. By common consent, rights in this property will be respected by officials of other agencies who in their turn will expect a similar courtesy.[25] In general, it is commonly admitted that the soil improvement associations of the TVA–extension service program belong to the county agent's office and will not be approached save through the good offices of the extension worker in charge. This is not a result of undemocratic attitudes but is simply because the associations were originally organized to serve certain administrative functions. An investment of time and energy was made, with the expectation that certain program responsibilities would be fulfilled. It would be idle to expect that the official in charge would be willing for an organization set up in response to his needs to be removed from his influence or diverted toward projects inimical to his purposes.

These considerations underline the coöptative character of the voluntary {234} groups organized to implement the test-demonstration program. Though not often formulated explicitly, the facts do not go unrecognized by the TVA staff. One participant put the matter in these terms, speaking of the early period:

> In the early days of the program, there was considerable shyness or aloofness on the part of farmers as to their willingness to test these experimental fertilizers. The seven land grant institutions of the Valley states agreed to contact individual farmers through their farm extension agents in the various counties. Many county agents when asked to organize the work in their particular counties accepted "all comers" as unit farm test demonstrators. During the first few years it appeared necessary for the county agent to run the show. The program was small and it could be efficiently operated by the county and assistant county agents through personal contacts with each individual farmer. The

county agent and the farmer made the farm plan and discussed as to how much and where the fertilizer was to be used and other improved practices. When the total amount of the material exceeded 30 tons or more, the county agent through the extension service ordered out the shipment. He made arrangements to pay the freight and when the car arrived he helped to load it on the test-demonstration farmer's conveyances. If any material was uncalled for he made the arrangements for its storage. In many instances he in turn collected from the individual farmer his proportion of the freight, paid the various bills and if a deficit existed, oft times made up the deficit from his personal funds or if there was a small surplus, put the surplus away or mingled it with his own funds. These efforts were truly coöperative ventures—one man one vote affairs—and the county agent was the one man one vote.

But as the program expanded, the necessity for functioning through groups became more pressing:

As the test-demonstration program caught on and neighbors heard about the "free TVA fertilizer," as it was often called, and about the results which were being obtained, many more farmers wanted to sign up. The introductory period of the program had passed and it now assumed a different aspect. Test-demonstration farmers began to convey a definite meaning, certain procedures were developing. Committees were set up for the purpose of administering the program. These committees were not particularly active at first, but at least they went through the motions of selecting new demonstrators and approving requisitions for phosphate, primarily upon the recommendation of the farm agent. Even at this stage the county agent still actually did the selecting but the responsibility for selection had the rubber stamp of approval of a local committee.

And we may note especially, in terms of the concept of coöptation:

In many ways the committee was convenient and useful as well as being a facade. Their names added prestige to the program. They often could handle complaints more tactfully and with better results than could the agent. They helped to publicize the activity and their farms—often already above the average farms—would be desirable to visit on their tours. Quite often the records of the committeemen were better kept; they had a better understanding of what was being attempted and as better farmers {235} were getting better results. Nothing succeeds like success and to the results obtained by many of these first test-demonstration committeemen the county agent could point with considerable pride.

Although these remarks pertain especially to the early years of the program, in fact the points adduced above were generally applicable to most of the associations as late as 1943.

Some officials of TVA headquarters look forward to the development of a strong coöperative movement based upon the network of soil improvement associations established under the test-demonstration program. Consequently, encouragement has been given to the local associations to branch out into marketing and purchasing activities in order that they might be strengthened by these added functions. The widening of activities would give the local farmers a greater stake in their associations and provide an impetus for incorporation and for otherwise improving their administration. Some accomplishments along this line have been noted: handling of lime by associations in Kentucky; distribution of the TVA fertilizer by-product, calcium silicate slag, through associations in northwestern Alabama; and purchasing and marketing ventures in northern Georgia through a regional federation embracing several counties.

On the whole, however, the state of the county associations in 1943 was not very satisfactory to those especially interested in the promotion of coöperatives. It was felt that the extension service remained hesitant about organizing unlimited membership organizations for fear that such organizations would "get out of their hands."[26] Of the 119 county associations in the Valley in 1943, a large proportion were thought to be more or less inactive, and their total net worth, less than $200,000, to be relatively small.

In addition to the county soil improvement associations, voluntary groupings at the community level have been organized. The watershed area projects,[27] another phase of the test-demonstration program, involve the distribution of fertilizer to entire farming communities. The formation of some community group, and the holding of at least a few initial meetings, is a condition of participation in the program. One of the tasks assigned to the organization is that of conducting a social and economic community self-survey, which theoretically is useful in setting a base line from which progress may be measured. Like the county associations, the community groups are formed at the initiative of the {236} extension-service workers and normally require considerable effort to be maintained.

Agricultural extension workers in the field consider the community organizations as effective means of furthering their work. Precisely because of their utility, and because of the large investment of time and energy required to activate them, there is a strong tendency to retain a considerable influence over the local groups. In fact, most of the groups are so weak, and require so much attention to be kept alive, that influence by the county agent's office is taken for granted. Investigation carried on in one Tennessee county—one of the better counties in the Valley—revealed that the assistant county agents assigned to the test-demonstration and readjustment programs were the mainstay of the community groups. One of these workers found it necessary to build up his community meetings by visiting key people several hours before the meeting in order to insure at least a minimal attendance. Conversations with local community leaders indicated that they did not think of the meetings as theirs, and came primarily for information, for phosphate clearance, and to some extent to please the extension agent. With some exceptions, there was observed a tendency for the associations to be primarily farmers' clubs, lacking links to other community groups.[28]

An assistant county agent in the county just referred to pointed out that the community meetings were sometimes mere formalities, required for the approval of farmers' requests to participate in the test-demonstration program. Such formal approval was required by TVA, and without such a requirement it is possible that fewer community meetings would have been held. Another agent in the same county said emphatically that he would not want the community organizations to become independent of his influence. While he viewed his role as primarily educational, nevertheless he felt that the continuation of a good program depended upon continued intervention and a measure of control on his part.

There is one generalization which emerges from a study of the TVA–extension service community organization work which helps to specify the conditions under which participation by local people is meaningful or is instead reduced to mere involvement. In many areas of the Tennessee {237} Valley traditional patterns of coöperation at the community level are only poorly developed, so that when new organizations are initiated they tend to remain external to the life of the community. When, however, the community possesses a history of coöperative effort, and a long acquaintance with its machinery, a new formal organization may be caught up within the older network and provided with a leadership and with well-established channels of access to the people of the area. This was notably true of a community in a mountain county of Tennessee which for many years had benefited from community organization work carried on by a Presbyterian home missionary and later by a pastor of some reputation. After a generation of such activity, the community was in a position to make the TVA fertilizer testing program an integral part of its established coöperative efforts. This meant that the community itself exercised a far stronger role than most *vis à vis* the extension worker. It may therefore be suggested as a hypothesis that the strength—in terms of local control and meaningful participation—of voluntary formal organizations established to implement administrative objectives will depend upon the initial social organization of the community.[29]

In the fertilizer distribution program, the voluntary associations as such do not appear to have operated significantly to build up loyalty to the TVA as an organization. The phosphate program has certainly been welcomed, and has created good feeling, but TVA has done as much as possible to play down its own role in favor of the extension service. Most farmers, however, are either aware of the TVA's role despite its lack of advertising or fail to distinguish the various brands of "government men." From the point of view of grass-roots attitudes and public support it seems fair to suggest that the readjustment program, which concentrates personnel and phosphates in areas most seriously affected by the impoundment of the reservoirs, does have a noticeable effect upon attitudes toward the Authority. Antagonism toward TVA at the grass roots often came from farmers whose lands were flooded and who had to move. But careful attention to the needs of those farmers over a period of years, and especially the concentration of advisory services and material aids in their direction, has mollified some of those early antagonisms. {238} However, it is not the associations as such which have accomplished that, though they probably do add a local flavor to the programs and provide a means for the involvement of leading individuals in their administration.

RURAL ELECTRIFICATION COÖPERATIVES

TVA distributes power at wholesale to municipalities, coöperatives, and large industrial contractors. The Authority's policy is to avoid direct distribution to the retail consumer wherever possible, preferring to sell to agencies of local government, such as municipal power boards, or to voluntary associations. The TVA Act of 1933 directed the agency to "give preference to States, counties, municipalities, and coöperative organizations of citizens or farmers, not organized or doing business for profit, but primarily for the purpose of supplying electricity to its own citizens or members."[30] In 1943 power was being sold to 83 municipalities and 45 coöperatives, as well as to such large industrial consumers as Monsanto Chemical Company and the Aluminum Company of America.

The organization of local associations for the distribution of TVA power to largely rural areas was begun in 1934. By 1939, contracts were in effect with 22 such groups, a figure which was more than doubled three years later. It is not contended by TVA that the coöperatives originated in a completely spontaneous movement, although there were always some local leaders who took the initiative. Speed was an important factor during the early period, not only because of pressure to get the program into operation to show some results, but because of the exigencies of controversy with the private power companies. "Spite lines" were being built,[31] and it was necessary to launch some of the associations very quickly in order to outmaneuver the power companies. This left little opportunity for full participation by the prospective membership, and little time for education.

Although the electrical coöperatives are independent organizations, relations to the Authority are very close. The contracts for sale of power {239} at wholesale are stringent, and include specification of rates at which the coöperative may resell the electricity at retail as well as stipulations with regard to finances and organization. Of course, TVA's punitive powers in case of infringements of the contracts are not very great in practice; the TVA would certainly hesitate to shut off electricity supplied to farmers simply because of the recalcitrance of officials of a coöperative. However, in the case of the coöperatives, the nature of the administrative relations makes for a large degree of influence on the part of TVA.[32] Many coöperatives are not strong enough to stand alone, and are dependent upon TVA for aid, as in setting up bookkeeping systems. Managers of the coöperatives are often selected jointly by TVA and REA.

The Authority inevitably accepts some public responsibility for the coöperatives, in terms of both financial success and public relations. It is therefore concerned that the coöperatives should take adequate measures to achieve success in both fields. Hence Authority personnel have initiated studies which might aid the coöperatives in maintaining their tax-exempt status; and officials of the TVA Personnel Department have attempted to apply some pressure to bring the coöperatives into line with TVA labor relations policy, especially as that relates to labor-management coöperation and the recognition of the International Brotherhood of Electrical Workers. But in facing these problems the Authority is involved in a dilemma: if it is actually to influence personnel policy and other phases of the organization of the coöperatives, it may be necessary to set up machinery for doing so; at the same time, to establish such machinery would strengthen Authority control and thus vitiate the original purposes for which the coöperative device was promoted. Circumstances have not demanded a choice between these alternatives, however, and the Authority has contented itself with informal controls beyond the terms of the contracts.

TVA officials carry on some educational work among the coöperatives, oriented in large part toward broadening the understanding of both officials and members in the principles of coöperative organization.[33] {240} There is said to be some difference in outlook within the Authority, however. Representatives of the Power Utilization Department reputedly interpret membership relations as public relations and are more interested in increasing electricity sales than in the development of Rochdale coöperatives.

Reverting again to the problem of approach to the voluntary association as an operational test of control, it appears that there is a tendency for representatives of other agencies to contact the electric coöperatives through TVA. However, while TVA would certainly be concerned about contacts inimical to its own good relations with the coöperatives, it does not exert the same close control over them in this situation as the county extension service agents do over the county soil improvement associations.

In general, people who buy electricity from local coöperatives in the Valley think of these organizations as "TVA co-ops." This is probably due in part to the wide publicity which attended the introduction of TVA and more particularly because, in some areas now serviced by coöperatives, TVA did sell directly to the consumer during the early period. Identification of the coöperatives with TVA is made easier by slight participation of the local people in the activities of the coöperatives. The members generally function as consumers of electricity and little more. Attendance at annual meetings is not large, and usually some inducement—refreshments, or perhaps a prize contest—is necessary. Most of the coöperatives require a five per cent attendance at annual meetings to constitute a quorum, but in 1942 one-fourth of the associations did not meet this requirement. Moreover, from the point of view of meaningful participation, annual meetings may not amount to much more than rallies. The monopoly character of electric power distribution and the dearth of personal contact between members and officials contribute to lack of participation.[34]

As might be expected, there is some tendency for the development of manager-director oligarchies within the coöperatives. Some managers {241} think of the members as customers, and many are reportedly director-controlled, with a tendency toward no contest at elections. The number of nominees at annual meetings will sometimes equal the number of directors, indicating that some prior decision has been made. There is also a tendency for some of the coöperatives to be run by their attorneys. Since many of the questions which come up are legal, the manager may call the attorney rather than the directors for advice and consultation. Thus a close relation between manager and attorney may grow up, with the latter exerting a strong informal influence.

A study of fifteen of the electrical coöperatives, concluded in 1939, made the following point:

> From a strictly coöperative point of view, the outstanding weakness of the associations studied is the lack of educational work among the membership. No association has an adequate educational program; all have dealt slightingly with coöperative membership education and membership relations. Association officers have not taken advantage of the opportunity to build up membership interest, loyalty, and a consciousness of ownership and pride among their members. If there were a fuller understanding among the membership of rural electrification through coöperatives, the load-building program would be easier and more successful. Unless more is done in the future than in the past on membership education, serious difficulties may arise which are likely to affect the actual existence of these associations as coöperative electricity-distributing units.

Here again we have a situation in which the pressure to get things done serves to reduce the organization from one having democratic implications to one which functions as a device for the accomplishment of administrative tasks.

It should be noted that precisely because the electrical coöperatives do have a definite leadership of their own; because they do have their own offices, hire their own employees, keep their own records;

because they do make significant decisions—for example, in regard to the extension of their lines; because they have relations with REA as well as TVA; for these reasons it is impossible to think of them as so weak and dependent as are the county soil improvement associations described above. In general, the electrical coöperatives, though not completely independent, are considered to be fairly strong. The net worth of the electrical coöperatives in 1943 was estimated at $3,000,000.

One comment which has been made on TVA's orientation toward coöperatives refers to what appears to be an exclusively rural perspective. The main group of coöperative specialists is bracketed within the Department of Agricultural Relations. The latter, being heavily committed to channeling all programs through the agricultural extension service, {242} exercises an inevitable inhibition on the field within which the coöperative specialists may work. The problem of ultimately overcoming the rural-urban distinction in administrative approach would seem to be especially pressing in relation to the electrical coöperatives. One of the basic over-all problems in the TVA power program has been to achieve a proper combination of territory in the distributing system so that urban areas might help support the cost of rural lines. The Authority, in deciding against a unified distributing system, compromised this issue at the outset, since presumably a unified system would have permitted big-city profits to be plowed into rural lines. Doubtless the political circumstances which shaped the inception of the program, and perhaps the intent of the Act, made that decision inescapable. But it seems apparent that administration of the program at the grass roots, through independent local agencies, has made it especially difficult for the Authority to take positive action on the question of persuading municipalities to extend their lines into rural areas.

OTHER VOLUNTARY ASSOCIATIONS

Two additional situations in which the Authority has sponsored voluntary associations as means of involving local people in the administration of its programs will be reviewed briefly.

Associations Handling Rental of TVA Lands.—The Authority controls large tracts of land adjacent to its reservoirs, amounting to almost 400,000 acres in 1943. Those portions of the land holdings considered suitable for agricultural use are leased to local farmers as part of the readjustment program. The program is channeled through the extension service; here again, as in fertilizer distribution, local associations make assignments of land and handle the money involved. In some areas, the same county soil conservation associations which handle the test-demonstration activities are utilized, but in others a new device has been introduced—a multiple-county association operating for an entire reservoir area: the Kentucky Reservoir Land-Use Association, the Norris Reservoir Area Land-Use Association, and the Chickamauga Land-Use Association. The associations classify the land—thus indirectly determining rental values—and receive and act on bids for licensing. Ten per cent of the licensing fees are retained by the associations to cover expenses.

The local organizations are not entirely independent, however. They receive guidance from extension service representatives, most of whom work out of the county agent's office with their salaries paid by TVA {243} under the readjustment program.[35] In effect, the associations are creatures of the extension service. They are not actually coöperatives, although that term appears in their charters, since the farmers who participate in the rental operations are not members. The Norris and Chickamauga associations' membership consists of the county agricultural planning committees of the constituent counties. But since the planning committees are actually sponsored and directed by the extension service,[36] the relation of the multiple-county organizations to the extension service remains the same. Contact with the associations is always through the extension service, a rule which holds—though not uniformly—for TVA officials as well as for outside agencies.

Audits conducted by the TVA Finance Department in 1940-1941 revealed certain discrepancies in the activities of the associations, such as leasing to local businessmen financially able to pay large rentals in advance and to employ farm labor to operate the land. To those making the audit, this seemed detrimental to the social objectives of the Authority's agricultural program. However, it is significant that in one Alabama county in which such leases were made, those with grass-roots perspectives justified the practice on the theory that leasing to absentee farmers did not alter the conditions which prevailed before the Authority bought the land. Since the local extension service personnel concurred in this judgment, we see here another example of the dilemma faced by TVA in turning over the administration of its program to grass-roots agencies. It is not known to what extent the eligibility of lessees for benefit payments under the Agricultural Adjustment Administration program has affected the licensing procedure. Sometimes AAA payments may have exceeded the cost of renting the land; it seems likely that under such circumstances there would be pressures which might result in depriving needy farmers of the use of the land.

In at least one of the reservoir areas, it was noted that there was considerable participation and independence on the part of community and county committeemen engaged in the leasing operations. Committeemen are paid by the day for their services, and the need for the land by the local population affords them opportunities for improving farm practices and for stimulating interest in rural planning.[37] {244}

Town Planning Commissions.—In 1938, a program of planning assistance to local communities was initiated by the Authority in coöperation with the State Planning Commissions of Alabama and Tennessee. This program is administered by an Urban Community Relations Division in the TVA's Regional Studies Department.[38] Two broad purposes are envisioned: (1) Execution of the Authority's responsibility for aiding in the readjustment of reservoir-affected communities. Impounding of large bodies of water behind the dams necessitates community readjustments—for example, to the creation of a new lake front—which should be properly integrated into a plan for the city or town as a whole. (2) The development of a project to demonstrate types of community action which will promote the general welfare: more specifically, to secure the establishment of planning as a recognized function of local government in the Valley.

Entering a situation in which the planning functions of local government were only poorly developed, the Authority has attempted to provide leadership in this direction by stimulating and aiding local planning programs of state planning commissions until public interest is developed sufficiently to insure the maintenance and expansion of the program without TVA sponsorship. TVA personnel also render direct aid by providing technical planning services to small communities, assistance in preparation of model planning legislation, and instructtional materials in planning for school curricula.

The TVA has applied its grass-roots concept in this field by working through agencies of the state governments and by supporting the creation of voluntary citizens' groups to serve as official planning commissions for local communities. However, it remains to the state agencies to guide the local commissions. The state commissions—only Alabama and Tennessee during the period up to 1943—may employ resident planning technicians whose salaries are reimbursed by TVA, or sometimes TVA will assign one of its own staff members to work under the commission. Activities are directed primarily toward the small communities which do not have resources to employ their own technicians and therefore must rely upon the state or TVA for this service. In 1943, {245} twenty-nine communities and planning commissions in Alabama and Tennessee were coöperating in the program. In reservoir-affected communities, the local commissions received guidance in waterfront development and readjustment, development of commercial, industrial, river terminal, and recreation facilities, and in comprehensive community planning and zoning; in war-affected communities, interest centered on programming and site selection of war housing, studies of school and recreation needs, and planning to relate emergency developments to long-time community needs.

The role of TVA in this activity is perhaps best defined by recalling that it has felt the need to stimulate local interest in planning. Consequently, though the program may be required by objective community needs, it can hardly be said to have been a response to needs felt at the grass roots. Authority personnel helped to draft enabling legislation for the states and organize the state planning machinery.[39] The relation between TVA officials and the technicians formally employed by the state commissions is very close. Indeed, there is some tendency for the latter to form a *de facto* field force of the TVA. If TVA has some particular need, it can operate through these men; thus, for practical purposes, there was reportedly little difference in the status of two planning technicians working in Tennessee, one of whom was a TVA employee, the other formally employed by the state commission under contract with TVA. In this program, the TVA assumes a definite leadership, which is inevitably reflected in informal organizational relations. This sometimes takes the form of a proprietary attitude toward the sponsored groups, as when the TVA planning officials refused permission to an independent investigator to interview members of local planning commissions, because of doubts and skepticism which might be raised in their minds by his questions. The need to go through channels—TVA or the state commissions or perhaps both—in contacting local groups is, in this case as in others we have mentioned, an indication of the coöptative character of the local groups. Having been set up for the accomplishment of certain program responsibilities, it is considered undesirable to place those objectives in jeopardy by permitting freedom of access by other individuals and agencies.[40] {246}

The organizational role of the local groups is emphasized further by the conscious choice made between two possible approaches: the planning personnel to be sponsored might be either a single planning engineer attached to an administrative office of the local government or it might be a planning commission. One of the reasons for preferring the latter was the belief that the commission would help to win public support for the planning program. In small communities especially, a nucleus of informed citizens might have considerable influence in this direction. Indeed, the public relations problem has supplied an underlying administrative rationale for the entire grass-roots approach in this field. Early in the program, it was found that TVA's attempt to work directly in community readjustment was not well received. The local citizens—or presumably some vocal portion of them—appear to have been suspicious of TVA because the same agency which was responsible for the damage was giving advice on readjustment to it. The device of working through state and local commissions, which could take responsibility for planning, was a convenient answer to this problem.[41]

---

[1] See above, p. 40.

[2] Yet this type of direct action on the part of other federal agencies, working out of Washington, has been roundly criticized by TVA officials.

[3] See above, pp. 13-14, and below, pp. 259-261.

[4] Perhaps as testimony to the effectiveness with which the grass-roots doctrine is circulated within the TVA, there appears to have developed a feeling that the Authority has somehow originated a unique administrative

device, binding the agency to its client public in some special way. This is partly referrable to enthusiasm, partly to the prevalent idea that other federal agencies, lacking the halo of regional decentralization, are unlikely to be really interested in democratic administration. It is hardly necessary to enter that controversy here, or to lay undue emphasis upon it. Yet although the grass-roots method is considered one of the major collateral objectives of the Authority, relatively little attention has been paid to the mechanics of its implementation and certainly the experience of other organizations facing the same problems and using voluntary associations has not been seriously studied inside the Authority.

[5] Karl Mannheim, *Man and Society in an Age of Reconstruction* (New York: Harcourt, Brace, 1941), pp. 44 ff.

[6] See José Ortega y Gasset, *The Revolt of the Masses* (New York: Norton, 1932).

[7] This is no necessary reflection on the integrity or the intentions of the responsible leadership. It is normal for programs infused with a moral content to be reduced to those elements in the program which are relevant to action. Thus the moral ideal of socialism has been reduced rather easily to concrete objectives, such as nationalization of industry. Administrative objectives, such as the establishment of a ramified system of citizens' committees, are similar.

[8] This distinction tends to melt away as the program administered comes closer to becoming an exclusive means of distributing the necessities of life, or if inducements are such as to eliminate any practical alternatives to participation.

[9] This may be the effective situation, even where there is not legal sponsorship. Thus, the local soil conservation districts are creatures of the state legislatures, but serviced by the Department of Agriculture's Soil Conservation Service. It is probably not inappropriately that they have been known in some areas as "SCS" districts.

[10] This might happen if the local groups formed an independent central organization, but that is not envisioned by the administrative agency, unless it already has control of a preexisting central organization, as when a national government utilizes a preexisting party structure to aid in the administration of its program.

[11] "A Theory of Agricultural Democracy," an address before the American Political Science Association, Chicago, December 28, 1940. Published as Extension Service Circular 355, March, 1941 (mim.). See also M. L. Wilson, *Democracy Has Boots* (New York: Carrick & Evans, 1939), chap. vii. Also, Howard R. Tolley, *The Farmer Citizen at War* (New York: The Macmillan Co., 1943), Pt. V.

[12] Wilson, "A Theory of Agricultural Democracy," p. 5. It is interesting to note that Mr. Wilson, Director of Extension Work, considered the AAA

program to have represented the practical beginning of agricultural democracy. The TVA agriculturists, loyal essentially to the local extension service organizations, would not have made such a statement.

[13] Carleton R. Ball, "Citizens Help Plan and Operate Action Programs," *Land Policy Review* (March-April, 1940), p. 19.

[14] See "The Land Use Planning Organization," County Planning Series No. 3, Bureau of Agricultural Economics, May, 1940; also, *Land Use Planning Under Way*, prepared by the BAE in coöperation with the Extension Service, FSA, SCA, AAA, and Forest Service, USDA, Washington, July, 1940; and John M. Gaus and Leon O. Wolcott, *Public Administration and the U. S. Department of Agriculture* (Chicago: Public Administration Service, 1940), pp. 151 ff.

[15] Reprinted as an appendix to Gaus and Wolcott, *op. cit.* Russell Lord (*The Agrarian Revival*, p. 193) notes extension service references to this agreement as "The Truce of Mt. Weather." Truce indeed, for by the middle of 1942 the Congress had scuttled the program, with the support of the American Farm Bureau Federation. See Charles M. Hardin, "The Bureau of Agricultural Economics Under Fire: A Study in Valuation Conflicts," *Journal of Farm Economics*, Vol. 28 (August, 1946).

[16] See John D. Lewis, "Democratic Planning in Agriculture," *American Political Science Review*, XXXV (April and June, 1941); also Neal C. Gross, "A Post Mortem on County Planning," *Journal of Farm Economics*, XXV (August, 1943); and Bryce Ryan, "Democratic Telesis and County Agricultural Planning," *Journal of Farm Economics*, XXII (November, 1940).

[17] Gross, p. 647.

[18] *Ibid.*, p. 653. Mr. Gross also points out that the units of planning tended to follow the convenience of the administrators, rather than local interest patterns.

[19] Lewis, *op. cit.*, p. 247.

[20] Unless "democracy" is reinterpreted, so that it reaches a higher level with the subordination of the mass to the organization. The above account, one sided in its emphasis, in no way deprecates the democratic aims of the initiators of the planning program. We are concerned here only with the explication of underlying trends to which the concept of coöptation lends significance.

[21] Organization of voluntary associations connected with electric power and fertilizer distribution also began early in the history of the Authority, but primarily as devices for the implementation of the major programs.

[22] The Act authorized the President to carry on this work, but this authority was delegated to TVA by Executive Order 6161, June, 1933.

[23] See above, pp. 130-132.

[24] Known variously as "soil erosion control," "soil improvement," or "soil conservation" associations.

[25] In December, 1942, the Forestry Relations Department of TVA explored the possibility of using county program planning committees in the development of Civilian Conservation Corps camp work loads. In reviewing the incident, a forestry official commented: "Since these committees are organizations sponsored and directed by the state extension services, it was deemed advisable to secure the consent of the state extension service in dealing directly with these organizations." The state extension director gave his permission, and stated that he would authorize his district agents to call meetings of county planning committees at any time such meetings were desired by the TVA foresters. "He indicated that original contacts should be made through the district agents, and that thereafter, meetings might be arranged for through the county agents." We see here an operational criterion for locating the *de facto* control of such associations. One function of the voluntary groups is to control the avenues of organized access to the farmers, thus strengthening the extension service by making it indispensable to other operating agencies.

[26] See above, p. 145, relative to the Farm Bureau's attitude toward coöperatives.

[27] See above, p. 132.

[28] Extension workers were often sincerely distressed by the weakness of the community organizations, and certainly were doing their best to strengthen them; but it appears that the pressure of responsibility for the administration of a definite program rather than for community organization work as such was often decisive. Lack of awareness of the elements of rural social organization may also have been a factor, in addition to the barrenness of initial social organization within some of the communities.

[29] Of course, initial social organization may also be a hindrance. If a new program is identified with existing organizations or leaders within the community, unintegrated elements may remain such and the community in its entirely will not be reached. Social structure offers a ready highway for effective administration, but its rigidities will usually require that some toll be paid. Moreover, to the extent that local control increases, the program of the administrator may be placed in jeopardy, for there is no guarantee that the vehicle he has used will not be taken out of his hands and all his work subverted.

[30] 48 Stat. 58 (1933), Sec. 10. Electrical coöperatives established under the TVA program are not essentially different from similar coöperatives set up in other parts of the country as a result of the work of Rural Electrification Administration. Indeed, the REA has also operated in the Valley, typically with a coöperative to which TVA functions as power wholesaler and REA as financing agency.

[31] The construction of lines into rural areas by private companies approximately paralleling lines proposed by a coöperative, thus undermining the possibility of successful financial operation by the coöperative. See "Rural Electrification Administration Activities in the Tennessee Valley Area," statement by W. E. Herring, Consulting Engineer, Rural Electrification Administration, *Reports and Exhibits of the Joint Committee Investigating the Tennessee Valley Authority*, Senate Document No. 56, Pt. 2, 76th Cong., 1st sess. (1939).

[32] This is apparently not true of the relations between TVA and the municipalities. Indeed, one observer reported in 1939 that there existed among TVA staff members "a decided dislike for the municipalities and a decided preference for the coöperatives." This was attributed to the feeling that "it is easier to deal with the co-ops because of the absence of politics, a greater willingness to accept TVA suggestions, and no desire on the part of the co-ops to make money for a general fund to reduce taxes." The relatively greater influence exercised by TVA over the coöperatives than over the municipalities was stressed by the manager of one of the large municipal power boards.

[33] Some of the educational work is simple trouble shooting, such as may be called for when there is a local movement in a small community for the establishment of an independent distributing system. TVA might consider such a proposal, if formulated, unfeasible; yet it would be embarrassing to have to refuse to sell electricity to a group which wanted it. In this connection, one may note that the Authority found itself in disagreement with REA as to the feasibility of certain proposed coöperatives in the Valley but was hardly in a position to deny electric service to several thousand farmers on that account.

[34] It seems doubtful that, during normal periods, the routine role of consumer represents an adequate basis for organizational participation. It may be questioned further whether it is always desirable to summon the participatory energies of a community without regard to the problems which present themselves to the people as real issues, or even to the limitations of such energies.

[35] There were 35 assistant county agents in "relocation, adjustment and land licensing" in 1943. These men also work on the test-demonstration program, which is intensified in the reservoir-affected counties.

[36] See above, p. 233, n. 25.

[37] The legal requirement that licensing be limited to periods of one year is noted by field workers as an obstacle to long-time planning, and this is probably one reason for the informal circumvention of competitive bidding for licenses.

[38] It was estimated that by 1943 the TVA had spent about $100,000 on its planning assistance program, which had stimulated the states to spend $25,000 and the local communities another $25,000. These are approxima-

tions, and represent five years of operation. The TVA figure does not include the salary of the chief of the Urban Community Relations Division, who has other responsibilities; it must also be noted that expenditures by the Authority are not strictly comparable to those by the states because some of the money represents time spent by TVA staff members on problems of interest peculiar to TVA.

[39] The Tennessee State Planning Commission is said to have been set up in the office of TVA's Regional Planning Director. It was formed from a committee established by the Tennessee legislature in 1933 to assist the work of the TVA.

[40] Another instance, this time involving a state commission rather than TVA, occurred when a representative of a private organization was rebuffed in his attempt to contact officials of a local community through the director of one of the state commissions. The director did not like the manner in which his assistance was sought and so refused to coöperate. Interesting here is that it appears to have been understood on all sides that the local groups should not be approached directly, but rather through the sponsoring agency.

[41] It is not contended that this is the only or the most important reason for the use of the grass-roots approach. Probably more significant is the professional interest on the part of planners to build up local understanding of and machinery for town planning. However, we are concerned to make explicit the underlying organizational needs which are served by the grass-roots policy.

Conclusion

GUIDING PRINCIPLES AND INTERPRETATION
A SUMMARY

CONCLUSION

GUIDING PRINCIPLES AND INTERPRETATION
A SUMMARY

The entire science considered as a body of formulae having coherent relations to one another is just a system of possible predicates—that is, of possible standpoints or methods to be employed in qualifying some particular experience whose nature or meaning is unclear to us.[1]

JOHN DEWEY

IT IS BELIEVED that the interpretation set forth in the preceding chapters provides a substantially correct picture of a significant aspect of the TVA's grass-roots policy at work. Far from remote, or divorced from what is considered pertinent by informed participants, the analysis reflects what is obvious to those who "know the score" in TVA.[2] Of course, this exposition is more explicit and systematic, and the relevant implications are more fully drawn out, but in main outline it can come as no surprise to leading officials of the Authority. This is not to suggest that there are no errors of detail, perhaps even of important detail. The nature of this kind of research precludes any full assurance on that. While much of the material is derived from documentary (though largely unpublished) sources, much is also based upon interviews with members of the organization and with those nonmembers who were in a position to be informed. Care was taken to rely upon only those who had an intimate, as opposed to hearsay, acquaintance with the events and personalities involved. Those who are familiar with the shadow-land of maneuver in large organizations will appreciate the difficulties, and the extent to which ultimate reliability depends upon the ability of the investigator to make the necessary discriminations. They will also recognize the need for insight and imagination if the significance of behavior, as it responds to structural constraints, is to be grasped. All this involves considerable risk.

If the use of personal interviews, gossip channels, working papers,[3] {250} and participation[4] opens the way for error, it remains, however, the only-way in which this type of sociological research can be carried on. A careful investigator can minimize error by such means as checking verbal statements against the documentary record, appraising the consistency of information supplied to him, and avoiding reliance on any single source. On the other hand, he will not restrict his data to that which is publicly acknowledged.

The possibilities of factual error, however great, are probably less important as hazard than the theoretical orientation of the study. To be sure, an empirical analysis of a particular organization, of its doctrine, of a phase of policy in action, of its interaction with other structures, was our objective. But in order to trace the dynamics of these events, it has been necessary to attempt a reconstruction, which is to say, a theory, of the conditions and forces which appear to have shaped the behavior of key participants.

Theoretical inquiry, when it is centered upon a particular historical structure or event, is always hazardous. This is due to the continuous tension between concern for a full grasp and interpretation of the materials under investigation as history, and special concern for the induction of abstract and general relations. Abstractions deal harshly with "the facts," choosing such emphases and highlighting such characteristics as may seem factitious, or at least distorted, to those who have a stake in an historically well-rounded apprehension of the events themselves. This is especially true in the analysis of individual personalities or social institutions, for these demand to be treated as wholes, with reference to their own central motives and purposes, rather than as occasions for the development of theoretical systems. This general, and perhaps inescapable, source of misunderstanding being admitted, let us review the concepts which have been used to order the materials of our inquiry.

SOCIOLOGICAL DIRECTIVES

This volume has been subtitled "A Study in the Sociology of Formal Organization." This means that the inquiry which it reports was shaped by sociological directives, more especially by a frame of reference for the theory of organization.[5] These directives are operationally relevant without, however, functioning as surrogates for inductive theory itself. That is, {251} while they provide criteria of significance, they do not tell us what is significant; while they provide tools for discrimination, they do not demand any special conclusions about the materials under investigation.[6] The fundamental elements of this frame of reference are these:

1. All formal organizations are molded by forces tangential to their rationally ordered structures and stated goals. Every formal organization—trade union, political party, army, corporation, etc.—attempts to mobilize human and technical resources as means for the achievement of its ends. However, the individuals within the system tend to resist being treated as means. They interact as wholes, bringing to bear their own special problems and purposes; moreover, the organization is imbedded in an institutional matrix and is therefore subject to pressures upon it from its environment, to which some general adjustment must be made. As a result, the organization may be significantly viewed as an adaptive social structure, facing problems which arise simply because it exists as an organization in an institu-

tional environment, independently of the special (economic, military, political) goals which called it into being.

2. It follows that there will develop an informal structure within the organization which will reflect the spontaneous efforts of individuals and subgroups to control the conditions of their existence. There will also develop informal lines of communication and control to and from other organizations within the environment. It is to these informal relations and structures that the attention of the sociologist will be primarily directed. He will look upon the formal structure, e.g., the official chain of command, as the special environment within and in relation to which the informal structure is built. He will search out the evolution of formal relations out of the informal ones.[7]

3. The informal structure will be at once indispensable to and consequential for the formal system of delegation and control itself. Wherever command over the responses of individuals is desired, some approach in terms of the spontaneous organization of loyalty and interest will be {252} necessary. In practice this means that the informal structure will be useful to the leadership and effective as a means of communication and persuasion. At the same time, it can be anticipated that some price will be paid in the shape of a distribution of power or adjustment of policy.

4. Adaptive social structures are to be analyzed in structural-functional terms.[8] This means that contemporary and variable behavior is related to a presumptively stable system of needs[9] and mechanisms. Every such structure has a set of basic needs and develops systematic means of self-defense. Observable organizational behavior is deemed explained within this frame of reference when it may be interpreted (and the interpretation confirmed) as a response to specified needs. "Where significant, the adaptation is dynamic in the sense that the utilization of self-defensive mechanisms results in structural transformations of the organization itself. The needs in question are organizational, not individual, and include: the security of the organization as a whole in relation to social forces in its environment; the stability of the lines of authority and communication; the stability of informal relations within the organization; the continuity of policy and of the sources of its determination; a homogeneity of outlook with respect to the meaning and role of the organization.

5. Analysis is directed to the internal relevance of organizational behavior. The execution of policy is viewed in terms of its effect upon the organization itself and its relations with others. This will tend to make the analysis inadequate as a report of program achievement, since that will be deemphasized in the interests of the purely organizational consequences of choice among alternatives in discretionary action.

6. Attention being focused on the structural conditions which influence behavior, we are directed to emphasize constraints, the limitation of alternatives imposed by the system upon its participants. This

will tend to give pessimistic overtones to the analysis, since such factors as good will and intelligence will be deemphasized.

7. As a consequence of the central status of constraint, tensions and dilemmas will be highlighted. Perhaps the most general source of tension {253} and paradox in this context may be expressed as the recalcitrance of the tools of action. Social action is always mediated by human structures, which generate new centers of need and power and interpose themselves between the actor and his goal. Commitments to others are indispensable in action: at the same time, the process of commitment results in tensions which have always to be overcome.

These principles define a frame of reference, a set of guiding ideas which at once justify and explain the kind of selection which the socioloogist will make in approaching organizational data. As we review some of the key concepts utilized in this study, the operational relevance of this frame of reference will be apparent.

UNANTICIPATED CONSEQUENCES IN ORGANIZED ACTION

The foregoing review of leading ideas directs our attention to the meaning of events. This leads us away from the problem of origins.[10] For the meaning of an act may be spelled out in its consequences, and these are not the same as the factors which called it into being. The meaning of any given administrative policy will thus require an excursion into the realm of its effects. These effects ramify widely, and those we select for study may not always seem relevant to the formal goals in terms of which the policy was established. Hence the search for meanings may seem to go rather far afield, from the viewpoint of those concerned only with the formal program. Any given event, such as the establishment of a large army cantonment, may have a multitude of effects in different directions: upon the economy of the area, upon the morals of its inhabitants, upon the pace of life, and so on. The freelance theorist may seek out the significance of the event in almost any set of consequences. But in accordance with the principle stated above, we may distinguish the random search for meanings—which can be, at one extreme, an aesthetic {254} interest—from the inquiry of the organizational analyst. The latter likewise selects consequences, but his frame of reference constrains his view: it is his task to trace such consequences as redound upon the organization in question; that is, such effects as have an internal relevance. Thus, only those consequences of the establishment of the army cantonment in a given area which result in adjustments of policy or structure in the administration of the cantonment will be relevant.

There is an obvious and familiar sense in which consequences are related to action: the articulation of means and ends demands that we weigh the consequences of alternative courses of action. Here consequences are anticipated. But it is a primary function of sociological inquiry to uncover systematically the sources of unanticipated consequence in purposive action.[11] This follows from the initial proposition in our frame of reference: "All formal organizations are molded by forces tangential to their rationally ordered structures and stated goals" (p. 251, above). Hence the notion of unanticipated consequence is a key analytical tool: where unintended effects occur, there is a presumption, though no assurance,[12] that sociologically identifiable forces are at work.

There are two logically fundamental sources of unanticipated consequence in social action, that is, two conditions which define the inherent predisposition for unanticipated consequences to occur:

1. *The limiting function of the end-in-view.*—A logically important but sociologically insignificant source of unanticipated consequence exists because the aim of action limits the perception of its ramified consequences.[13] This is legitimate and necessary, for not all consequences are relevant to the aim. But here there arises a persistent dilemma. This very necessity to "keep your eye on the ball"—which demands the construction of a rational system explicitly relating means and ends—will {255} restrain the actor from taking account of those consequences which indirectly shape the means and ends of policy. Because of the necessarily abstract and selective character of the formal criteria of judgment, there will always be a minimum residue of unanticipated consequence.[14]

2. *Commitment as a basic mechanism in the generation of unanticipated consequences.*—The sociologically significant source of unanticipated consequences inherent in the organizational process may be summed up in the concept of "commitment." This term has been used throughout this study to focus attention upon the structural conditions which shape organizational behavior. This is in line with the sociological directive, stated above, that constraints imposed by the system will be emphasized. A commitment in social action is an enforced line of action; it refers to decision dictated by the force of circumstance with the result that the free or scientific adjustment of means and ends is effectively limited. The commitment may be to goals, as where the existence of an organization in relation to a client public depends on the fulfillment of certain objectives;[15] or, less obviously, to means, derived from the recalcitrant nature of the tools at hand. The commitments generated by the use of self-activating and recalcitrant tools are expressed in the proliferation of unintended consequences.[16] {256}

The types of commitment in organizational behavior identify the conditions under which a high frequency of unanticipated consequences may be expected to occur:

i) *Commitments enforced by uniquely organizational imperatives.*—An organizational system, whatever the need or intent which called it into being, generates imperatives derived from the need to maintain the system. We can say that once having taken the organizational road we are committed to action which will fulfill the requirements of order, discipline, unity, defense, and consent. These imperatives may demand measures of adaptation unforeseen by the initiators of the action, and may, indeed, result in a deflection of their original goals. Thus the tendency to work toward organizational unity will commit the organization as a whole to a policy originally relevant to only a part of the program. This becomes especially true where a unifying doctrine is given definite content by one subgroup: in order to preserve its special interpretation the subgroup presses for the extension of that interpretation to the entire organization so that the special content may be institutionalized.[17]

ii) *Commitments enforced by the social character of the personnel.*—The human tools of action come to an organization shaped in special but systematic ways. Levels of aspiration and training, social ideals, class interest—these and similar factors will have molded the character of the personnel. This will make staff members resistant to demands which are inconsistent with their accustomed views and habits; the freedom of choice of the employer will be restricted, and he will find it necessary in some measure to conform to the received views and habits of the personnel. Thus, in recruiting, failure to take into account initial commitments induced by special social origins will create a situation favorable to the generation of unanticipated consequences. The TVA's agricultural leadership brought with it ideological and organizational commitments which influenced over-all policy. This was a basically uncontrolled element in the organization. It is noteworthy that where the character of any organization is self-consciously controlled, recruitment is rigidly qualified by the criterion of social (class, familial, racial) origin.

iii) *Commitments enforced by institutionalization.*—Because organizations are social systems, goals or procedures tend to achieve an established, value-impregnated status. We say that they become institutionalized. {257} Commitment to established patterns is generated, thus again restricting choice and enforcing special lines of conduct. The attempt to commit an organization to some course of action utilizes this principle when it emphasizes the creation of an established policy, or other forms of precedent. Further, the tendency of established relations and procedures to persist and extend themselves, will create the unintended consequence of committing the organization to greater involvement than provided for in the initial decision to act.[18] Where policy becomes institutionalized as doctrine, unanalyzed elements will persist, and effective behavior will be framed in terms of immediate necessities. An official doctrine whose terms are not operationally relevant will be given content in action, but this content will be informed by the special interests and problems of those to whom delegation is made. Hence doctrinal formulations will tend to reinforce the inherent hazard of delegation.[19] A variation of this situation occurs when the role of participants comes to overshadow in importance the achievement of formal goals. Action then becomes irresponsible, with respect to the formal goals, as in the "fanatical" behavior of the TVA agriculturists.[20]

iv) *Commitments enforced by the social and cultural environment.*—Any attempt to intervene in history will, if it is to do more than comment upon events, find it necessary to conform to some general restraints imposed from without. The organizers of this attempt are committed to using forms of intervention consistent with the going social structure and cultural patterns. Those who ascend to power must face a host of received problems; shifts in public opinion will demand the reformulation of doctrine; the rise of competing organizations will have to be faced; and so on. The institutional context of organizational decision, when not taken into account, will result in unanticipated consequences. Thus intervention in a situation charged with conflict will mean that contending forces will weigh the consequences of that intervention for their own battle lines. The intervening organization must therefore qualify decision in terms of an outside controversy into which it is drawn despite itself. More obviously, the existence of centers of power and interest in the social environment will set up resistances to, or accept and shape to some degree, the program of the organization.

v) *Commitments enforced by the centers of interest generated in the {258} course of action.*—The organizational process continuously generates subordinate and allied groupings whose leaderships come to have a stake in the organizational status quo. This generation of centers of interest is inherent in the act of delegation. The latter derives its precarious quality from the necessity to permit discretion in the execution of function or command. But in the exercise of discretion there is a tendency for decisions to be qualified by the special goals and problems of those to whom delegation is made. Moreover, in the discretionary behavior of a section of the apparatus, action is taken in the name of the organization as a whole; the latter may then be committed to a policy or course of action which was not anticipated by its formal program. In other words, the lack of effective control over the tangential informal goals of individuals and subgroups within an organization tends to divert it from its initial path. This holds true whether delegation is to members and parts of a single organization, or to other organizations, as in the TVA's relation to the land-grant colleges.

These types of commitment create persistent tensions or dilemmas.[21] In a sense, they set the problems of decision and control, for we have identified here the key points at which organizational control breaks down. Operationally, a breakdown of control is evidenced in the generation of observable unanticipated consequences. This is the same as to say that significant possibilities inherent in the situation have not been taken into account. The extension of control, with concomitant minimization of unintended consequence, is achieved as and if the frame of reference for theory and action points the way to the significant forces at work.

The problems indicated here are perennial because they reflect the interplay of more or less irreconcilable commitments: to the goals and needs of the organization and at the same time to the special demands of the tools or means at hand. Commitment to the tools of action is indispensable; it is of the nature of these tools to be dynamic and self-activating; yet the pursuit of the goals which initiated action demands continuous effort to control the instruments it has generated. This is a general source of tension in all action mediated by human, and especially organizational, tools.

The systematized commitments of an organization define its character. Day-to-day decision, relevant to the actual problems met in the translation of policy into action, create precedents, alliances, effective symbols, and personal loyalties which transform the organization from a profane, {259} manipulable instrument into something having a sacred status and thus resistant to treatment simply as a means to some external goal. That is why organizations are often cast aside when new goals are sought.

The analysis of commitment is thus an effective tool for making explicit the structural factors relevant to decision in organized action. Attention is directed to the concrete process of choice, selecting those factors in the environment of decision which limit alternatives and enforce uniformities of behavior. "When we ask, "To what are we committed?" we are speaking of the logic of action, not of contractual obligations freely assumed. So long as goals are given, and the impulse to act persists, there will be a series of enforced lines of action demanded by the nature of the tools at hand. These commitments may lead to unanticipated consequences resulting in a deflection of original goals.[22]

THE COÖPTATIVE MECHANISM

The frame of reference stated above includes the directive that organizational behavior be analyzed in terms of organizational response to organizational need. One such need is specified as "the security of the organization as a whole in relation to social forces in its environment." Responses, moreover, are themselves repetitive—may be thought of as mechanisms, following the terminology of analytical psychology in its analysis of the ego and its mechanisms of defense. One such organizational mechanism is ideology; another, which has been the primary focus of this study, we have termed coöptation. In the Introduction we have defined this concept as "the process of absorbing new elements into the leadership or policy-determining structure of an organization as a means of averting threats to its stability or existence." Further, this general mechanism assumes two basic forms: Formal coöptation, when there is a need to establish the legitimacy of authority or the administrative accessibility of the relevant public; and informal coöptation, when there is a need of adjustment to the pressure of specific centers of power within the community.

Coöptation in administration is a process whereby either power or the burdens of power, or both, are shared. On the one hand, the actual center of authority and decision may be shifted or made more inclusive, with or without any public recognition of the change; on the other hand, {260} public responsibility for and participation in the exercise of authority may be shared with new elements, with or without the actual redistribution of power itself. The organizational imperatives which define the need for coöptation arise out of a situation in which formal authority is actually or potentially in a state of imbalance with respect to its institutional environment. On the one hand, the formal authority may fail to reflect the true balance of power within the community; on the other hand, it may lack a sense of historical legitimacy, or be unable to mobilize the community for action. Failure to reflect the true balance of power will necessitate a realistic adjustment to those centers of institutional strength which are in a position to strike organized blows and thus to enforce concrete demands. This issue may be met by the kind of coöptation which results in an actual sharing of power. However, the need for a sense of legitimacy may require an adjustment to the people in their undifferentiated aspect, in order that a feeling of general acceptance may be developed. For this purpose, it may not be necessary actually to share power: the creation of a "front" or the open incorporation of accepted elements into the structure of the organization may suffice. In this way, an aura of respectability will be gradually transferred from the coöpted elements to the organization as a whole, and at the same time a vehicle of administrative accessibility may be established.

We may suggest the hypothesis: Coöptation which results in an actual sharing of power will tend to operate informally, and correlatively, coöptation oriented toward legitimization or accessibility will tend to be effected through formal devices. Thus, an opposition party may be formally coöpted into a political administration through such a device as the appointment of opposition leaders to ministerial posts. This device may be utilized when an actual sharing of power is envisioned, but it is especially useful when its object is the creation of public solidarity, the legitimization of the representativeness of the government. In such circumstances, the opposition leaders may become the prisoners of the government, exchanging the hope of future power (through achieving public credit for holding office in a time of crisis) for the present function of sharing responsibility for the acts of the administration. The formal, public character of the coöptation is essential to the end in view. On the other hand, when coöptation is to fulfill the function of an adjustment to organized centers of institutional power within the community, it may be necessary to maintain relationships which, however consequential, are informal and covert. If adjustment to specific nucleuses of {261} power becomes public, then the legitimacy of the formal authority, as representative of a theoretically undifferentiated community (the "people as a whole"), may be undermined. It therefore becomes useful and often essential for such coöptation to remain in the shadowland of informal interaction.

The informal coöptation of existing nucleuses of power into the total (formal plus informal) policy-determining structure of an organization, symptomatic of an underlying stress, is a mechanism of adjustment to concrete forces. On this level, interaction occurs among those who are in a position to muster forces and make them count, which means that the stake is a substantive reallocation of authority, rather than any purely verbal readjustment. Formal coöptation, however, is rather more ambiguous in relation to *de facto* reallocations of power. The sense of insecurity which is interpreted by a leadership as indicating a need for an increased sense of legitimacy in the community is a response to something generalized and diffuse. There is no hard-headed demand for a sharing of power coming from self-conscious institutions which are in a position to challenge the formal authority itself. The way things seem becomes, in this context, more important than the way they are, with the result that verbal formulas (degenerating readily into propaganda), and formal organizational devices, appear to be adequate to fill the need. The problem becomes one of manipulating public opinion, something which is necessarily beside the point when dealing with an organized interest group having an established and self-conscious leadership.

Formal coöptation ostensibly shares authority, but in doing so is involved in a dilemma. The real point is the sharing of the public symbols or administrative burdens of authority, and consequently public responsibility, without the transfer of substantive power; it therefore becomes necessary to insure that the coöpted elements do not get out of hand, do not take advantage of their formal position to encroach upon the actual arena of decision. Consequently, formal coöptation requires informal control over the coöpted elements lest the unity of command and decision be imperiled. This paradox is one of the sources of persistent tension between theory and practice in organizational behavior. The leadership, by the very nature of its position, is committed to two conflicting goals: if it ignores the need for participation, the goal of coöperation may be jeopardized; if participation is allowed to go too far, the continuity of leadership and policy may be threatened.[23] {262}

THE EMPIRICAL ARGUMENT RESTATED

Apart from the interest of analytical theory, the statement above explains the special focus of this inquiry and the basis for its obviously selective approach to the TVA experience. That frame of reference has guided the empirical analysis, of which the following is a brief recapitulation.

1. *The grass-roots theory became a protective ideology.*—In chapter ii, an attempt was made to explain the high self-consciousness of the TVA, as expressed in the grass-roots doctrine, on the basis of the function of that doctrine in facilitating acceptance of the Authority in its area of operation and in fulfilling the need for some general justification of its existence as a unique type of governmental agency. The TVA was revolutionary both to the attitudes of local people and institutions and to the federal governmental system. By adopting the grass-roots doctrine the Authority was able to stand as the champion of local institutions and at the same time to devise a point of view which could be utilized in general justification of its managerial autonomy within the federal system. However, allegiance to this doctrine, and translation of it into policy commitments, have created serious disaffection between TVA and other branches of the federal government, including the Department of Agriculture and the Department of the Interior. As a result, on the basis of the TVA experience, these departments have been moved to oppose the extension of the TVA form of organization to other areas, a fact which is consequential for the future of the Authority itself.

2. *The agricultural program was delegated[24] to an organized administrative constituency.*—In the major example within TVA of grass-roots procedure—the Authority's fertilizer distribution program—there was constructed a strong constituency-relation involving the land-grant college system on the one hand and the Agricultural Relations Department of TVA on the other. This constituency relation may be viewed as a case of informal coöptation, wherein strong centers of influence in {263} the Valley were absorbed covertly into the policy-determining structure of the TVA. The TVA's Agricultural Relations Department assumed a definite character, including a set of sentiments valuing the land-grant college system as such and accepting the mission of defending that system within the Authority. In effecting this representation, the TVA agriculturists have been able to take advantage of the special prerogatives accruing to them from their formal status as an integral part of the Authority, including the exercise of discretion within their own assigned jurisdiction and the exertion of pressure upon the evolution of general policy within the Authority as a whole. The special role and character of the TVA agricultural group limited its outlook with respect to the participation of Negro institutions as grass-roots resources and created a special relation to the American Farm Bureau Federation. Yet the operation of this coöptative process probably did much to enhance the stability of the TVA within its area and especially to make possible the mobilization of support in an hour of need. In this sense, one cannot speak of the decisions which led to this situation as mistakes.

3. *In a context of controversy, the TVA's commitments to its agricultural constituency resulted in a factional alignment involving unanticipated consequences for its role on the national scene.*—In the exercise of discretion in agriculture, the TVA entered a situation charged with organizational and political conflict. The New Deal agricultural agencies, such as Farm Security Administration and Soil Conservation Service, came under attack of the powerful American Farm Bureau Federation, which thought of them as threats to its special avenue of access to the farm population, the extension services of the land-grant colleges. Under the pressure of its agriculturists, the Authority did not recognize Farm Security Administration and sought to exclude Soil Conservation Service from operation within the Valley area. This resulted in the politically paradoxical situation that the eminently New Deal TVA failed to support agencies with which it shared a political communion, and aligned itself with the enemies of those agencies.

4. *Under the pressure of its agriculturists, the TVA gradually altered a significant aspect of its character as a conservation agency.*—The TVA agricultural group, reflecting local attitudes and interests, fought against the policy of utilizing public ownership of land as a conservation measure and thus effectively contributed to the alteration of the initial policy of the Authority in this respect. The issue of public ownership is taken as character-defining in the sense that it is a focus of controversy and division, and it was such within the TVA for an extended period. The {264} single-minded pursuit of its ideological and constituency interests led the agricultural group to involve the Authority in a controversy with the U. S. Department of the Interior over the management of TVA-owned lands. In this matter, as in those mentioned above, the mechanics of pressure and representation are detailed in the body of the study.

5. *The grass-roots utilization of voluntary associations represents a sharing of the burdens of and responsibility for power, rather than of power itself.*—In chapter vii, the voluntary association device—especially, but not exclusively, in the agricultural program—is interpreted as a case of formal coöptation, primarily for promoting organized access to the public but also as a means of supporting the legitimacy of the TVA program. Typically, this has meant that actual authority, and to a large extent the organizational machinery, has been retained in the hands of the administering agency. After nine years of operation, the county soil associations handling TVA fertilizer were found to be still tools of the county agent system, to which the TVA test-demonstration program was delegated. In connection with this analysis, an operational test for locating control over coöpted citizens' groups is described, as suggested in the question: Is approach to the association by outside elements channeled through officials of the coöpting agency?

IMPLICATIONS FOR DEMOCRATIC PLANNING

In venturing this interpretation, we are pointing to some of the significant problems which must be faced when the attempt is made to combine democracy and planning.[25] For planning implies large-scale intervention and the extended use of organizational instruments. If such concepts as "democracy" are to be more than honorific symbols which mobilize opinion, it is essential to make explicit the forces which will operate to qualify and perhaps transform the democratic process.

There are three major considerations, three sources of paradox and tension, which emerge out of this study, so far as it has implications for democratic planning:

1. Ideologies must be seen in the context of the needs they serve; it is a reliable assumption that something more urgent than patterns of belief will lie behind the strong advocacy of doctrine by an organizational leadership. The unanalyzed terms must be closely examined, for it is also a reliable expectation that some covert adaptation in terms of immediate necessity will have provided content for emotion-laden but {265} procedurally indefinite terms. It is perhaps a good rule to see whether the ambiguities, the inherent dilemmas, of the attempt to fulfill doctrinal demands are squarely faced; where these are denied out of hand, the ideological function of the doctrine is obliquely confirmed.

2. It is essential to recognize that power in a community is distributed among those who can mobilize resources—organizational, psychological, and economic—and these can effectively shape the character and role of governmental instrumentalities. This has a dual significance. It may result in the perversion of policy determined through representative institutions; and at the same time, this fact offers a tool for ensuring the responsibility of public agencies to their client publics. Consequently, it is naive to suppose that there is anything inherently bad in the situation wherein private organizations paralleling but independent of a governmental administrative structure have a decisive influence on its social policy. Again, however, the situation is inherently ambiguous. This ambiguity must be explicitly recognized, and its mechanics understood, if realistic controls are to be instituted.

3. The tendency of democratic participation to break down into administrative involvement requires continuous attention. This must be seen as part of the organizational problem of democracy and not as a matter of the morals or good will of administrative agents. A realistic examination of the factors which define formal coöptation also permits us to make explicit precisely those points at which changes in procedure will be effective in reinforcing meaningful democracy.

In general, therefore, we have been concerned to formulate some of the underlying tendencies which are likely to inhibit the democratic process. Like all conservative or pessimistic criticism, such a statement of inherent problems seems to cast doubt upon the possibility of complete democratic achievement. It does cast such a doubt. The alternative, however, is the transformation of democracy into a Utopian notion which, unaware of its internal dangers, is unarmed to meet them.

The TVA has been a particularly good subject for the analysis of these problems. This is so precisely because it may be said that the Authority has, on the whole, very effectively achieved some of its major purposes, including the mobilization of a staff of very high quality. No one is surprised when a weak or corrupt governmental agency does not fulfill its doctrinal promise. When, however, a morally strong and fundamentally honest organization is subject to the kind of process we have described, then the pervasive significance of that process becomes materially enhanced. In a sense, it is just because TVA is a relatively good {266} example of democratic administration that the evidences of weakness in this respect are so important. It is just because the TVA stands as something of a shining example of incorruptibility in such major matters as noncapitulation to local political interests in the hiring of personnel or to local utility interests in public power policy that the evidence of covert coöptation in the agricultural program attains its general significance.

For the things which are important in the analysis of democracy are those which bind the hands of good men. We then learn that something more than virtue is necessary in the realm of circumstance and power.

---

[1] John Dewey, *Problems of Men* (New York: Philosophical Library, 1946), p. 221.

[2] Although responsibility for the analysis rests solely with the author, it should be emphasized that this study was made possible by the willingness of TVA to make its records and personnel available. This is a happy precedent which we may hope will be followed by other organizations, public and private.

[3] Some of the materials quoted in the study are unofficial in the sense that they would be vigorously edited before receiving even the public status of a memorandum sent to another department within TVA. This would be so with comparable documents in any large organization, public or private.

[4] The author spent most of his year's stay at TVA in daily contact with personnel of the agency. A number of weeks was spent in intensive contact with extension service personnel in the field.

[5] For a fuller statement than the summary which follows see Philip Selznick, "Foundations of the Theory of Organization," *American Sociological Review*, XIII (February, 1948).

[6] Thus, while approaching his materials within a guiding frame of reference, the author was not committed by this framework to any special hypothesis about the actual events. Indeed, he began his work with the hypothesis that informally the grass-roots policy would mean domination by TVA, because of its resources, energy, and program. After the first two months in the field, however, this hypothesis was abandoned as a major illuminating notion.

[7] For discussion of informal organization, see P. J. Roethlisberger and W. J. Dickson, *Management and the Worker* (Cambridge: Harvard University Press, 1941), pp. 524 ff; also Chester I. Barnard, *The Functions of the Executive* (Cambridge: Harvard University Press, 1938), chap. ix; Wilbert E. Moore, *Industrial Relations and the Social Order* (New York: The Macmillan Co., 1946), chap. xv.

[8] See Talcott Parsons, "The Present Position and Prospects of Systematic Theory in Sociology," in George Gurvitch and Wilbert E. Moore (eds.), *Twentieth Century Sociology* (New York: Philosophical Library, 1945).

[9] As Robert K. Merton has pointed out to the author, the concept of "basic needs" in organizational analysis may be open to objections similar to those against the concept of instinct. To be sure, the needs require independent demonstration; they should be theoretically grounded independently of imputations from observed responses. However, we may use the notion of "organizational need" if we understand that it refers to stable systems of variables which, with respect to many changes in organizational structure and behavior, are independent.

[10] In terms of origins, the TVA's policy—though not the grass-roots doctrine *qua* doctrine—of channeling its agricultural program through the land-grant colleges of the Valley states may be adequately referred to such factors as the nature of the formal agricultural program, the resources available for its implementation, and the administrative rationale which seemed conclusive to leading participants. Moreover, these factors may sustain the continued existence of the policy, and it may therefore seem superfluous when extraneous factors are brought in and somewhat tangential explanations are offered. But when we direct our attention to the meaning of the policy in terms of certain indirect but internally relevant consequences—as for the role of TVA in the agricultural controversy,—we have begun to recast our observation of the policy (taken as a set of events) itself. We are then concerned not with the question, "how did the grass-roots policy come into being?" but with the question, "what are the implications of the grass-roots policy for the organizational position and character of TVA?"

[11] Consequences unanticipated from the viewpoint of the formal structure are not necessarily undesired. On the contrary, the result may be a satisfactory adjustment to internal and external circumstances, upon which the leadership may find it convenient to declare that the results were actually intended, though close analysis might show that this is actually a ration-

alization. In this type of unintended consequence, some need is fulfilled. The same unintended consequence may fulfill a need for a part of the organization and at the same time cause difficulties for the whole, and conversely. Many unintended consequences are, of course, sociologically irrelevant. For an early statement of this general problem, see Robert K. Merton, "The Unanticipated Consequences of Purposive Social Action," *American Sociological Review*, Vol. I (December, 1936).

[12] Where unintended consequences occur due to error, or to individual idiosyncrasy, they are sociologically irrelevant. However, there is often, though not always, a systematically nonrational factor at work whose presence is manifested by mistakes and personality problems.

[13] This follows, of course, from the hypothetical, and therefore discriminating and ordering, status of the end-in-view. See John Dewey, *Logic: The Theory of Inquiry* (New York: Henry Holt, 1938), pp. 496–497.

[14] The use of the terms "end-in-view" and "anticipated" may easily lead to the fallacy of formulating this problem as one of the subjective awareness of the participants. This is a serious error. What is really involved is that which is anticipated or unanticipated by the system of discrimination and judgment which is applied to the means at hand. This may, and very often does, involve subjective anticipation or its want, but need not do so. Moreover, the system may be adjusted so as to be able to take account of factors previously unpredicted and uncontrolled. This addition of systematically formulated criteria of relevance occurs continuously, as in the recognition of morale factors in industry. In the situation detailed above, the high self-consciousness of the American Farm Bureau Federation apparently led it to anticipate the possible rivalry from a new organization set up under the Agricultural Adjustment Administration, since it took steps to ward off this threat. See above, p. 161. This is no accidental perspicacity but a result of the systematic consideration of just such possible consequences from the implementation of new legislation. However, the tendency to ignore factors not considered by the formal system—not so much subjectively as in regard to the competence of the system to control them—is inherent in the necessities of action and can never be eliminated.

[15] As in the TVA's commitment to become a successful electric power business; this type of commitment was much milder in the distribution of fertilizer, permitting adaptation in this field which would contribute to the fulfillment of the prior commitment to electricity.

[16] Our use of the notion of unanticipated consequence assumes that the functional significance of such consequences is traceable within a specific field of influence and interaction. Thus price decisions made by a small enterprise affect the market (cumulatively with others), with ultimate unanticipated and uncontrolled consequence for future pricing decision. This is not an organizational process. When, however, the retailer builds up good will or makes decisions which will enforce his dependence upon some manufacturer, these are organizational acts within a theoretically

controllable field, and are analyzable within the frame of reference set forth above.

[17] In the TVA, the agriculturists made vigorous efforts to extend their interpretation of the grass-roots policy to the Authority as a whole; in respect to the federal government, the TVA attempts to have its special interpretation of administrative decentralization become general public policy.

[18] See above, p. 70 f.

[19] We have reviewed above, pp. 59–64, the unanalyzed abstractions in TVA's grass-roots doctrine, which are given content and meaning by the pressure of urgent organizational imperatives.

[20] See above, pp. 205 ff.

[21] In effect, we have restated here some of the basic points made in the discussion above of the inherent dilemmas of the TVA doctrine. See pp. 69–74.

[22] The British Labor Party, when it assumed power in 1945, had to accept a large number of commitments which followed simply from the effort to govern in those circumstances, independently of its special program. "Meeting a crisis," in a women's club as well as in a cabinet, is a precondition for the institution of special measures. To assume leadership is to accept these conditions.

[23] The analysis of unanticipated consequence and commitment is indispensable to the interpretation of behavior in terms of the coöptative mechanism. The commitments made in the course of action generate unanticipated consequences; in analyzing the function of these consequences we must construct a theory which will explain them as events consistent with the needs and potentialities of the system. At the same time, it must be understood that to formulate such defensive mechanisms as coöptation is to state possible predicates. For the full understanding of organization it will be necessary to construct a system of such relevant responses which can serve to illuminate concrete cases.

[24] Some TVA officials would question the use of "delegated" here. However, this seems to be the most significant summary word to use, in terms of its implications. Moreover, in his own summation of TVA policy upon the occasion of his leaving the TVA chairmanship, David E. Lilienthal said: "The TVA has by persistent effort delegated and thereby decentralized its functions...." New York *Times*, November 13, 1946, p. 56.

[25] We speak here only of the implications of this analysis. That the grass-roots approach has wider implications of importance for democratic planning is sufficiently obvious.

PUBLISHED SOURCES

Anderson, W. A. "Farm Youth in the 4-H Club," Department of Rural Sociology Mimeograph Bulletin No. 14, Cornell University Agricultural Experiment Station, Ithaca, N.Y. (May, 1944)

Association of Land-Grant Colleges and Universities. *Proceedings of the 48th Annual Convention*, Washington, D.C. (1934).

Baker, Gladys. *The County Agent* (Chicago: University of Chicago Press, 1939).

Ball, Carleton R. "Citizens Help Plan and Operate Action Programs," *Land Policy Review* (March–April, 1940).

———. *A Study of the Work of the Land-Grant Colleges in the Tennessee Valley Area in Coöperation with the Tennessee Valley Authority*, prepared under the auspices of the Coördinating Committee of the USDA, the TVA, and Valley States Land-Grant Colleges (No imprint, 1939).

Barnard, C. I. *Dilemmas of Leadership in the Democratic Process* (Princeton: Princeton University Press, 1939).

Barnard, C. I. *The Functions of the Executive* (Cambridge: Harvard University Press, 1938).

Davidson, Donald. "Political Regionalism and Administrative Regionalism," *Annals of the American Academy of Political and Social Science*, Vol. 207 (January, 1941).

Gaus, J. M. "A Theory of Organization in Public Administration," *The Frontiers of Public Administration* (Chicago: University of Chicago Press, 1936).

Gaus, J. M., and L. D. Wolcott. *Public Administration and the U. S. Department of Agriculture* (Chicago: Public Administration Service, 1940).

Gerth, H. H., and C. Wright Mills. *From Max Weber* (New York: Oxford University Press, 1946).

Gibson, D. L. "The Clientele of the Agricultural Extension Service," *Michigan Agricultural Experiment Station Quarterly Bulletin*, Vol. 1 (May, 1944).

Gray, L. C. "Our Land Policy To-day," *Land Policy Review*, Vol. 1 (May-June, 1938).

Gregory, Clifford V. "The American Farm Bureau Federation and the AAA," *Annals of the American Academy of Political and Social Science*, Vol. 179 (May, 1935).

Gross, Neal C. "A Post Mortem on County Planning," *Journal of Farm Economics*, Vol. XXV (August, 1943).

Gulick, Luther, and Lydall Urwick (eds.). *Papers on the Science of Administration* (New York: Institute of Public Administration, Columbia University, 1937).

Hardin, Charles M. "The Bureau of Agricultural Economics Under Fire, A Study in Valuation Conflicts," *Journal of Farm Economics*, Vol. 28 (August, 1946).

Hartford, Ellis F. *Our Common Mooring* (Athens: University of Georgia Press, 1941).

Hodgkin, Carlyle. "Agriculture's Got Troubles," a series in *Successful Farming* (March, April, and May, 1943).

Hoffer, C. R. "Selected Social Factors Affecting Participation in Agricultural Extension Work," Michigan State College Agricultural Experiment Station Special Bulletin 331 (June, 1944).

Iowa State College. *The Role of the Land-Grant College in Governmental Agricultural Programs* (Ames: Iowa State College Bulletin, June, 1938).

Key, V. O., Jr. *The Administration of Federal Grants to the States* (Chicago: Public Administration Service, 1937).

Key, V. O., Jr. "Politics and Administration," in L. D. White (ed.), *The Future of Government in the U. S.* (Chicago: University of Chicago Press, 1942).

Leiserson, Avery. *Administrative Regulation: A Study in the Representation of Interests* (Chicago: University of Chicago Press, 1942).

Lewis, John D. "Democratic Planning in Agriculture," *American Political Science Review*, Vol. XXV (April and June, 1941).

Lilienthal, David E. "Management—Responsible or Dominant?," *Public Administration Review*, Vol. I (Summer, 1941).

——. "The TVA: An Experiment in the Grass-Roots Administration of Federal Functions," Address before the Southern Political Science Association, Knoxville, Tennessee, November 10, 1939 (TVA Information Office: no date).

——. *TVA: Democracy on the March* (New York: Pocket Books, 1945).

Lilienthal, David E., and Robert H. Marquis. "The Conduct of Business Enterprises by the Federal Government," *Harvard Law Review*, Vol. 54 (February, 1941).

Lord, Russell. *The Agrarian Revival* (New York: American Association for Adult Education, 1939).

Malinowski, Bronislaw. *The Dynamics of Culture Change* (New Haven: Yale University Press, 1945).

——. *A Scientific Theory of Culture* (Chapel Hill: University of North Carolina Press, 1944).

Merton, Robert K. "The Unanticipated Consequences of Purposive Social Action," *American Sociological Review*, Vol. I (December, 1936).

Mooney, James D., and Alan C. Reiley. *The Principles of Organization* (New York: Harpers, 1939).

New Horizons in Public Administration. A Symposium (University, Alabama: University of Alabama Press, 1945).

Parsons, Talcott. "The Present Position and Prospects of Systematic Theory in Sociology," in Georges Gurvitch and Wilbert E. Moore (eds.), *Twentieth Century Sociology* (New York: Philosophical Library, 1945).

——. *The Structure of Social Action* (New York: McGraw-Hill, 1937).

Pincus, William. "Shall We Have More TVA's?," *Public Administration Review*, Vol. V (Spring, 1945).

Pritchett, C. Herman. *The Tennessee Valley Authority: A Study in Public Administration* (Chapel Hill: University of North Carolina Press, 1943).

Richards, E. C. M. "The Future of TVA Forestry," *Journal of Forestry*, Vol. 36 (July, 1938).

Ryan, Bryce. "Democratic Telesis and County Agricultural Planning," *Journal of Farm Economics*, Vol. XXII (November, 1940).

Selznick, Philip. "Foundations of the Theory of Organization," *American Sociological Review*, Vol. XIII (February, 1948).

Stephens, Oren. "FSA Fights for Its Life," *Harper's* (April, 1943).

Tolley, Howard R. *The Farmer Citizen at War* (New York: Macmillan, 1943).

U. S. Bureau of Agricultural Economics. "The Land Use Planning Organization," County Planning Series No. 3 (May, 1940).

——. *Land Use Planning Under Way* (Washington, D.C.: Government Printing Office, July, 1940).

U. S. Congress. House. Committee on Appropriations. *Hearings* on Agriculture Department Appropriation Bill for 1944, 78th Cong., 1st sess. (1943).

U. S. Congress. House. Committee on Military Affairs. *Muscle Shoals Hearings*, 72d Cong., lst sess. (1932).

——. Joint Committee on Reduction of Nonessential Federal Expenditures. *Hearings*, 77th Cong., 3d sess. (1942).

——. Joint Committee to Investigate the Tennessee Valley Authority. *Hearings*, 75th Cong., 3d sess. (1938). *Report*, 76th Cong., 1st sess. (1939).

U. S. Congress. Senate. Committee on Agriculture and Forestry. *Hearings* on S. 2555, "To provide for the creation of conservation authorities ...," 75th Cong., 1st sess. (1937).

——. Senate. Committee on Agriculture and Forestry. *Hearings* on S. 2361, "To amend the TVA Act of 1933, as amended, with respect to the manner of the exercise of the power of condemnation by the TVA and to require the receipts of the Authority to be covered into the Treasury," 77th Cong., 2d sess. (1942).

——. Senate. Committee on Agriculture and Forestry. *Hearings* on S. 1251, "To provide for the establishment of a national soil fertility policy and program ...," 80th Cong., 1st sess. (1947).

——. Senate. Committee on Civil Service. *Hearings* on H.R. 960, "An Act Extending the Classified Civil Service," 76th Cong., 3d sess. (1940).

——. Senate. Committee on Commerce. *Hearings* on S. 555, "A Bill to Create a Missouri Valley Authority," 79th Cong., 1st sess. (1945).

——. Senate. Committee on Irrigation and Reclamation. *Hearings* on S. 555, "A Bill of Create a Missouri Valley Authority," 79th Cong., 1st sess. (1945).

U. S. National Resources Planning Board. *Public Land Acquisition in a National Land-Use Program*, Pt. I: *Rural Lands* (Washington, D.C.: Government Printing Office, 1940).

Urwick, Lydall. *The Elements of Administration* (New York: Harpers, 1943).

Wilson, M. C. "How and to What Extent is the Extension Service Reaching Low-Income Farm Families?" USDA Extension Service Circular 375 (December, 1941).

Works, G. A., and Barton Morgan. *The Land Grant Colleges*, Staff Study No. 10, Advisory Committee on Education (Washington, D.C.: Government Printing Office, 1939)

.

INDEX

Page numbers reference the prior editions, for continuity purposes. This pagination is inserted into the present text within {brackets}. The Index tracks these numbers.

ABOUT THE AUTHOR

Philip Selznick (1919-2010) was Professor of Law and Sociology at the University of California, Berkeley. He was the founding chair of the Center for the Study of Law and Society and of the Jurisprudence and Social Policy Program in the School of Law. His other acclaimed books include *The Organizational Weapon; Leadership in Administration; Law, Society and Industrial Justice; Law and Society in Transition* (with Philippe Nonet); *The Moral Commonwealth; The Communitarian Persuasion*; and, in 2008, *A Humanist Science.*

Visit us at *www.quidprobooks.com*.

Made in the USA
Charleston, SC
08 June 2012